Nosson —
removal of Beacon Hill Reservoir

Empire of Water

Empire of Water

An Environmental and Political History
of the New York City Water Supply

DAVID SOLL

CORNELL UNIVERSITY PRESS

Ithaca & London

First published 2013 by Cornell University Press

Printed in the United States of America

Library of Congress Cataloging-in-Publication Data

Soll, David, 1971–

  Empire of water: an environmental and political history of the New York City water supply / David Soll.

     p. cm.

  Includes bibliographical references and index.

  ISBN 978-0-8014-4990-1 (cloth: alk. paper)

  1. Water-supply—Environmental aspects—New York (State)—New York—History.   2. Water-supply—Political aspects—New York (State)—New York—History.   3. Watershed management—New York (State)—New York—History.   4. Water-supply engineering—New York (State)—New York—History.   I. Title.

  TD225.N5S65 2013

  363.6'1097471—dc23   2012033792

Cornell University Press strives to use environmentally responsible suppliers and materials to the fullest extent possible in the publishing of its books. Such materials include vegetable-based, low-VOC inks and acid-free papers that are recycled, totally chlorine-free, or partly composed of nonwood fibers. For further information, visit our website at www.cornellpress.cornell.edu.

Cloth printing   10 9 8 7 6 5 4 3 2 1

Contents

Preface

While I was writing this book, several people inquired about my motivations for exploring the history of New York City's water supply. I never managed to answer the question to my complete satisfaction, nor, I imagine, to theirs. Nonetheless, I am certain about one thing: uncovering the intricacies of government—its many moving parts, diverse personalities, and dueling interests—made this an especially fascinating topic to research.

One of the first documents that I came across in my research was a song by engineers of the Board of Water Supply, the entity the city established to construct a new water system in the early twentieth century. I soon realized that these songs (it turned out that there were dozens of them) were invaluable sources because they were written by and for board employees and as such offered unique insights into their values and desires. Government was not simply a bunch of faceless bureaucrats; it was also engineers toiling in the Catskill Mountains who longed for "Broadway with all its sights."

Throughout this book, I frequently refer to the "city" or to various government agencies. Nonetheless, government is not monolithic, and I have tried to highlight the fissures that divide most public bodies. I also focused on particular individuals who played an outsize role in shaping public decisions. This is a story of laws, courts, and legislatures, but it is also a story of individuals, public and private, who challenged government to act in the public interest.

Many people lent a hand in my efforts to trace the journeys of the individuals and governments who influenced the development of the city's water system. My first debt is to the writers who came before me. There are several full-length works on New York City's water supply; in addition, many authors have devoted portions of books to the city's water network. Diane Galusha's *Liquid Assets*, and *Water-Works*, a beautifully produced book by Kevin Bone and Gina Pollara, provided both a detailed summary of the city's water supply history and important factual details. Many scholars cannot avail themselves of such comprehensive overviews of their topic.

Shortly after embarking on my research, I encountered a significant stumbling block—because of post-9/11 security concerns, I was not permitted to use the archives of the city's Department of Environmental Protection, the primary municipal repository of information on the water network. At the time, I worried that the absence of such sources would make it difficult to tell the environmental history of the water system in rich detail. Fortunately, these concerns proved unfounded, in large part due to the abundance of materials stored in various archives and libraries.

I appreciate the support and guidance of the New York State Archives, where I served as a Larry J. Hackman Fellow in 2008. James Folts and his staff helped me locate useful materials. Their dedication to preserving state records is a model of public service. I would also like to thank James Stimpert and staff members in special collections at the Eisenhower Library at Johns Hopkins University, where I conducted research in the Abel Wolman Papers. The papers are a priceless storehouse of information on water supply development and sanitary engineering in the United States. They provided a fascinating behind-the-scenes perspective on the city's water supply decisions in the 1950s. I am also grateful to the Rockefeller Archive Center for its support.

The libraries and archives of the Catskill Mountain region were indispensable. Their newspaper clipping files, tourism brochures, and other materials greatly enriched my knowledge and have (hopefully) enlivened this book. The records of commission hearings for reservoir damage cases maintained by Ulster County (the Ashokan Reservoir Archives) and Hartwick College (the Board of Water Supply Transcripts) are a historian's delight, providing information on many aspects of the waterworks expansion experience. Sylvia Rozelle, town clerk of Olive, generously gave me access to the town's rich collection of documents related to the construction of the Ashokan Reservoir. Thanks also to Edythe Ann Quinn at Hartwick College, who took an interest in my project and generously shared materials she collected from Catskill publications.

The book's final two chapters draw heavily on personal interviews. Virginia Scheer and Nancy Burnett deserve enormous credit for having the foresight and dedication to record interviews with most of the key figures in the 1997 Watershed Memorandum of Agreement negotiations. These interviews, aptly titled *Behind the Scenes*, greatly enriched my understanding of this critical period in the city's water supply history. They also enabled me to conduct more rewarding interviews with several of these same individuals.

My understanding of developments since the signing of the Watershed Memorandum of Agreement (1997–present) was decisively shaped by the extensive interviews I conducted with public employees, individuals employed by nongovernmental organizations active in New York City water supply issues, and activists. A complete list of interviewees appears in the bibliography. Warren Liebold of the New York City Department of Environmental Protection and Tara Collins of the Watershed Agricultural Council were particularly generous with their time.

I was fortunate to have the opportunity to work with Michael McGandy and Sarah Grossman at Cornell University Press. Michael's consummate professionalism and expert guidance were the perfect fit for a first-time author. I also appreciate the helpful feedback of Martin Melosi and an anonymous reviewer on an earlier draft of this book.

I was particularly fortunate to have many scholars to consult as I researched and wrote this book. David Stradling took an aspiring environmental historian under his wing and offered excellent guidance. His writings on the connections between New York City's water system and Catskill residents were an inspiration. I greatly appreciate the time and intelligence he contributed to this book. Brian Donahue offered a vigorous blend of scholarship and activism. His support, editorial suggestions, and wide range of contacts greatly improved this book. Jane Kamensky's keen sense of narrative and advice on a wide range of personal and professional topics were an invaluable resource.

An Andrew W. Mellon Postdoctoral Fellowship in Environmental Studies at Lafayette College provided the time and resources to revise the manuscript and conduct additional research. The foundation has contributed enormously to my development and that of many young scholars throughout the country. I also greatly appreciate the Mellon Foundation's financial support for this book. The fellowship enabled me to take up residence on the banks of the Delaware River while I revised the manuscript. For that and many other reasons, Lafayette was the perfect place to think about my research in the larger context of environmental studies. Dru Germanoski, who led the college's environmental initiative, was a consistent source of encouragement. Thanks also to two of my biggest supporters at Lafayette, D. C. Jackson and Elisabeth Rosen. Your blend of criticism and encouragement was the perfect recipe.

Friends and family have been a critical source of support over the years. Howard Hacker, Paul Skeith, and Jeff Gold provided delightful hiking respites in a place far from New York City. Martina and Bernd Leger are wonderful friends who unfailingly supported me in my quest, even though it meant leaving Massachusetts. Wanda Fleck and Alan and Lois Palestine provided comfortable homes away from home on research trips to the New York City area. They may not have always been interested in the details, but the kindness of my mother, Susan Soll, and sisters, Lisa and Nancy Soll, helped sustain me. Although my father passed away before I began this project, his analytical mind-set and respect for hard work and scholarly rigor were never far from my mind. I am confident he would have enjoyed reading this book and arguing its finer points with me.

Finally, I am incredibly indebted to my wife, Sara Wise, and son Maxwell. Despite the obvious financial downside, Sara encouraged me to return to school. She is the ideal editor: firm but not inflexible, generous with praise when warranted, but not afraid to deliver bad news. Her patience and diligent editing greatly improved the book. Most authors do not have the luxury of a professional in-house editor; I consider myself especially fortunate. Most important, her love and sense of humor helped me persevere when the pages came too slowly. Max, you are the joy of my life. You grew up with this project, and your smiling face and good nature helped me keep it in perspective. Not surprisingly, you asked more questions about it than anyone else. Thanks for actually caring about the answers.

Empire of Water

The Evolution of a Water System

Road atlas designers generally divide New York State into two sections. The densely populated southeastern quadrant consumes a page, while the sprawling northern and western sections require two pages. This convention highlights a rather startling reality: with the exception of the Hudson River, the most prominent inland bodies of water in the region that stretches from Long Island to the Catskill Mountains are the reservoirs that serve New York City. North of the city, in Westchester and Putnam Counties, the dozen reservoirs and three controlled lakes of the Croton water system dot the landscape. A hundred miles to the northwest, in the Catskill Mountains, a different picture emerges. In a region laced with rivers, creeks, and streams but bereft of large lakes, six substantial reservoirs dominate their valleys, serene repositories for the snowmelt and water that courses down the hillsides.

The construction and management of New York City's water supply in the twentieth century is the subject of this book. New York City began designing its Catskill Mountains water network in 1905. It completed the Cannonsville Reservoir, the final component of its mountain water system, in the mid-1960s. Damming local streams and two major tributaries of the Delaware River provided New York with an enormous volume of water, enough to meet 90 percent of its needs.[1] *Empire of Water* tells the story of this 90 percent, and the challenge of operating and maintaining one of the world's most extensive water networks.

The regional implications of New York City's waterworks expansion were profound. The flooding of rural communities by enormous reservoirs and the resulting economic dislocation were predictable consequences of waterworks construction. But focusing exclusively on the remaking of the rural landscape obscures the ways in which the desire for water also transformed the suburban and urban recreational and cultural landscape.[2] Exploring developments through this regional prism reveals the diverse and often surprising effects of waterworks expansion

Fig. 1. New York City watershed. New York constructed its water system over the course of 130 years, beginning in Westchester County in 1837 and ultimately ending in the western Catskills in 1967. (Collection, New York City Department of Environmental Protection. Courtesy of the New York City Department of Environmental Protection)

on city, suburb, and countryside. Some communities disappeared; others saw familiar streams altered beyond recognition; others acquired new parks. The construction of the city's water system reconfigured the natural and built environments of southeastern New York State, from Long Island to the headwaters of the Delaware River, 125 miles northwest of the city. New York City's "hydrological commons," the area affected by its perpetual pursuit of more water, was both much larger and more intricately constructed than historians have recognized.[3]

The eighteenth-century residents of a city notorious for its unsavory water could scarcely have envisioned the elaborate supply and distribution network that the municipal government would construct in the nineteenth and twentieth centuries. Colonial-era New Yorkers relied on hundreds of wells that tapped a network of underground streams. When British troops occupied New York during the Revolutionary War, they destroyed an incipient project to construct a reservoir at the outskirts of the city. (This was the first serious attempt to construct a public water supply.) Independence did not bring an improved water supply. Instead, a combination of rapid population growth and poor sanitation practices led to a steady deterioration of water quality, especially for working-class New Yorkers, who could not afford the high prices charged for more pristine supplies. It was not until the 1830s, in the wake of a ferocious cholera epidemic that killed more than three thousand residents, that New Yorkers voted to build a public water network to convey high-quality supplies from outside the city.[4]

The Croton water system—so named because it drew on the waters of the Croton River and its tributaries—was one of the city's signal accomplishments of the nineteenth century. New York harnessed the labor of Irish immigrants and the capital of America's most prosperous port to build a substantial reservoir and a forty-one-mile aqueduct to carry its waters to hundreds of thousands of citizens.[5] In addition to altering the landscape of northern Westchester County, construction of the Croton system reshaped the city itself. Workers built the elegant High Bridge, which carried the Croton Aqueduct over the Harlem River and into the city. The aqueduct terminated at the Yorkville Receiving Reservoir, which would become a prominent Central Park landmark. From there, water flowed to the distributing reservoir at Murray Hill, a hulking four-story water tank that dominated the surrounding countryside. New Yorkers greeted the arrival of Croton water with unbridled enthusiasm.[6] Municipal officials even commissioned a poem to mark the occasion. George Pope Morris's verse clearly conveyed the joy and wonder of urban dwellers who had long abided inadequate and impure supplies: "Water shouts a glad hosanna! / Bubbles up the earth to bless! / Cheers it like the precious manna / In the barren wilderness."[7]

This enthusiasm proved short-lived. Within a decade, the inability of the water system to keep pace with increasing consumption had become apparent, prompting calls for its expansion. By the 1850s, two patterns that would define New York City's approach to water supply for the next century had emerged: the city would seek pure water from well beyond its borders, and it would engage in a perpetual struggle to secure enough water to meet the demands of the metropolis. Although these trends began in the nineteenth century, they became more pronounced in the twentieth century, when the scale of New York City's waterworks projects and the level of demand for fresh water reached unprecedented heights. By the time it was finally completed in 1911, the Croton system bore little resemblance to its first incarnation; its original reservoir and aqueduct had been supplanted by larger versions, and its reservoirs and lakes extended over two upstate counties.[8] Generations of engineers and workers had toiled to collect virtually every drop of water from the river and its tributaries, but the Croton's modest yield could not meet the seemingly insatiable demand for water.

Increasing demand for water reflected political and demographic changes. The consolidation of Brooklyn, Manhattan, and other communities into Greater New York in 1898 almost doubled the city's population, placing enormous strains on the water network. Manhattan soon discovered the inadequacy of the water systems it had inherited; locating and tapping new water sources to supply the outer boroughs became one of the first tests of the city's commitment to Brooklyn and the other newly annexed communities. Even as they continued to expand the Croton system, municipal officials recognized the need for a much larger source to accommodate the demands of rapid population growth and increasing per capita water consumption. They again eyed the region north of the city, but this time they looked west of the Hudson, to the dense network of streams that laced the Catskill Mountains, roughly one hundred miles from New York. Like their counterparts in Los Angeles, San Francisco, and Boston, New York's political leaders envisioned a long-distance water delivery system that would draw on rural resources to meet the growing demand of urban dwellers.[9]

The near-continuous process of system expansion from the 1830s to the 1960s shaped and reshaped New York City and its suburbs. The original Croton system relied on reservoirs and pumping stations located within the city to deliver water to urban households. Modernization of the Croton system and the widespread introduction of water from the Catskills beginning in 1917 rendered obsolete most of the water infrastructure within city limits. New York City "recycled" these parcels, greatly enriching the cultural and recreational landscape of the city and its suburbs. Two of Manhattan's most prominent landmarks—the New York Public Library and Central Park's Great Lawn—occupy former reservoir sites. On Long Island, property acquired by Brooklyn to protect its water supply became the nucleus of the island's parkway and state park systems. In Westchester County, the state converted the path of the Old Croton Aqueduct into a linear park, which has become one of the most beloved and utilized recreational resources in suburban New York.[10]

The tendency of historical accounts to focus almost exclusively on the technical aspects of reservoir and aqueduct construction has consigned these urban and suburban by-products of rural waterworks expansion to the realm of anecdote and curiosity. This represents a missed opportunity to connect water supply expansion to larger themes of public space and the churning of the built environment. By taking a regional perspective on New York City's water system, this book illustrates the wide-ranging and enduring significance of these lesser-known products of waterworks expansion.

In addition to widening the geographical scope of inquiry, I explore developments over a long stretch of time—from the twilight of the nineteenth century to the early years of the twenty-first century. The ecological effects of water development did not end with the completion of a dam. On the contrary, it was only when a reservoir went into service that scientists and watershed residents began to fully appreciate the wide-ranging environmental consequences. Similarly, political tensions between watershed residents and city officials did not disappear with the end of active construction. Frustration with New York City's post-construction watershed policies—its poor record of road maintenance, inconsistent releases of water from its reservoirs back into streams and rivers, restrictions on recreational use of its watershed properties, and frequent challenges of local property tax assessments—explains much of the resentment felt by Catskill residents.[11] The city's management of its hydraulic network often proved just as controversial as its decision to tap Catskill waters in the first place.

This broader approach—exploring the life of a water network, rather than simply its birth—highlights the connections between politics and ecology that lie at the heart of this book. As its title suggests, New York City enjoyed immense autonomy in designing, constructing, and managing its water network. This urban dominance began in 1905, when the city received permission from the state to divert Catskill streams, and continued into the 1960s, when New York completed its last mountain reservoir. Upstate communities surrendered farms, homesteads, and hamlets to the needs of the downstate metropolis. Waterworks development also altered the recreational landscape of the Catskills. Large releases of water from the reservoirs changed the flow, character, and even the temperature of mountain streams. Catskill residents were prohibited from hiking and hunting on thousands of acres of reservoir buffer lands. Until the 1970s, watershed residents won only marginal victories in their attempts to resist New York City's incursions. Their appeals to the state to block reservoir construction amounted to pro forma exercises in free speech; the era of the environmental impact statement had not yet arrived. Those who lost their homes to new reservoirs or saw their businesses fade with the disruption of community living patterns generally received some compensation from the special boards the state created to hear their claims. However, some residents entitled to monetary awards received nothing or next to nothing. In theory, watershed expansion took place under the watchful eye of state and federal authorities charged with balancing the interests of country and city. In practice, the scales of power were rigged in favor of New York City.

The increasing influence of ecology and the erosion of urban political clout led to a gradual but tectonic shift in the balance of power between the city and watershed residents. By the 1970s, the empire that had constructed the vast water network was crumbling. New York City teetered on the edge of bankruptcy, and hundreds of thousands of urban dwellers had fled the city in search of a better life. The state government that intervened to save the city's finances also began to take its environmental responsibilities more seriously, no longer reflexively supporting New York's management of its water system. New York City constructed its water network in the pre-ecological era, but it was compelled to change the way it operated this system to reflect evolving ecological priorities and knowledge. This transformation in environmental governance, dramatic as it was, did not lead to imperial withdrawal. New York City continued to tap the region's streams and rivers to provide water for residents in the city and northern suburbs. Ironically, the emphasis on watershed protection that emerged in the 1990s enmeshed New York City more tightly into the fabric of daily life in the Catskills. But the nature of the relationship between watershed residents and the city had changed dramatically since the 1970s.

The Watershed Memorandum of Agreement, generally known as the MOA, was the most significant evidence of change. The MOA, signed in 1997, balanced the city's desire to minimize human activity in the watershed with the recognition that economic development and expanded recreational opportunities were critical to watershed residents' quality of life. New York secured permission to acquire more land in the areas that supplied its reservoirs. In exchange it agreed to invest tens of millions of dollars into local economies, open up more of its watershed holdings to hunting and hiking, and fund a wide variety of projects to help local farmers, residents, and communities improve water quality. Although the MOA did not eliminate disputes between New York City and watershed communities, it signaled the end of the imperial era. It created new institutional mechanisms designed to forestall serious disputes, and firmly bound the fate of the Catskills to New York City's water supply.[12]

The MOA was based on the concept of ecosystem services—the recognition that preserving natural processes such as pollination and water filtration can be a cost-effective means of achieving environmental goals. A participant in the watershed negotiations likened the plight of Catskill residents to that of Amazon natives: "It behooves the rest of the world to provide some sort of economic alternatives to destroying the rain forest. Well, the same is true up here. The city should provide an economic alternative."[13] The success of the watershed negotiations demonstrated that ecosystem services, a strategy hitherto employed almost exclusively in remote regions, could help protect the economy and ecology of developed areas as well. The New York City watershed agreement is recognized as an international model of environmental dispute resolution and water management. Experts from around the world have visited the watershed to learn more about its programs, and it has informed the resolution of environmental conflicts in distant regions.[14]

In the American context, the MOA represented an important shift in environmental management. The combination of increasingly sophisticated understanding of ecological processes and more flexible governing regimes gave rise to a more collaborative mode of environmental politics by the late twentieth century. For most of the century, New York City officials maintained a static conception of natural processes; as long as the city continued to draw its water from relatively undeveloped mountain watersheds, its main concern was quantity, not quality. The transition from a command-and-control style of environmental politics to a more cooperative approach that recognized the need to collaborate with watershed residents did not happen overnight. Well into the 1990s, municipal officials clung to the old verities and attempted to unilaterally impose a system of centralized ecological management on its watersheds. Resistance from Catskill residents and the spur of extraordinarily expensive federal water quality statutes were the proximate causes of New York's decision to embrace a new mode of environmental governance. Nonetheless, the arrival of a new breed of environmentally minded city leaders with a more holistic vision of watershed management was critical in forging a collaborative partnership with watershed residents. The ability of these partners to deliver clean water to more than nine million people without sacrificing rural economic vitality ranks as one of the most significant American environmental success stories of the last thirty years.[15]

The evolution of New York City's water supply system reflects significant changes in environmental policy and thought in twentieth-century America. It embodies both the conservationist urge to use nature to meet human needs, and the preservationist impulse to minimize human interference in natural processes. Perhaps most important, it casts the environmental revolution of the final decades of the century in a new light. Most accounts of late twentieth-century American environmental policy emphasize the role of federal statutes and the legal disputes they spawned in establishing the parameters of environmental change and reform.[16] It would be foolhardy to deny the powerful influence of federal regulations. In the case of New York City's water supply, they spurred the negotiations that produced the MOA. But this narrative needs revising. Businesses, environmental organizations, and average citizens did more than fight to restrict or expand the reach of federal regulations in court. As the New York City watershed negotiations reveal, they sometimes worked together to reconcile conflicting social, economic, and environmental goals. Regulations may have established the frame of possible outcomes, but citizens, government officials, and other parties filled in the important details.[17]

Two aspects of the transformation in environmental governance—water conservation and watershed recreation—receive particular attention in this book. They are each important to the environmental history of the water system, and both topics underscore the regional dimensions of water supply expansion. New York bears the dubious distinction of being the last major American city to install water meters in all residences. It did not complete the job until the 1990s, decades after most other cities began using consumption as the basis for water charges. The reluctance to install meters reflected the overall laxity toward water conservation that prevailed until the 1980s. The ability to continually expand the water supply discouraged the development of meaningful conservation policies. Leaking sinks and toilets in Manhattan bore testament to the abundance of Catskill water. The absence of residential metering and low water charges led to some of the highest per capita rates of water consumption in the United States. New Yorkers significantly reduced consumption when droughts threatened to lead to water shortages, but usage shot back up when the rains returned and the reservoirs filled.

The city finally broke this cycle in the 1990s, when it launched the nation's largest toilet replacement program. Despite the first marked increase in New York City's population in decades, overall water usage decreased significantly. Although largely obscured by the MOA, the reduction in water consumption represented a clear break with decades of intransigence and ended all speculation about further expansion of the water system. In 1987, water expert Edwin Clark observed that New York City "has the reputation for the best-engineered and worst-managed water system in the nation."[18] By the late 1990s, this charge no longer rang true. New York was slowly learning how to share regional resources with its neighbors.

One of the most prized resources in the watersheds was recreational space. The development of New York's supply network reshaped recreation on both land and water. Reservoirs provided new fishing opportunities, but water releases from these reservoirs also altered stream conditions, creating challenges for those who sought to swim and fish in the Catskills, the birthplace of American fly-fishing. Conflicts over Catskill waters in the 1970s foreshadowed disputes about recreational access to city-owned lands in the 1990s. To enhance the protection of its water sources, New York City began to purchase land throughout the Catskills, eventually acquiring tens of thousands of acres of mountain holdings. These acquisitions threatened to severely limit access to parcels that had formerly been available for hunting, hiking, and other forms of recreation. Reconciling public health and the desire for recreational access to newly acquired properties loomed as a major challenge.

City and state officials and local residents have worked diligently to expand recreational access to New York's expanding watershed holdings since the signing of the MOA. This collaborative process has revealed areas of common ground between the needs of the water system and residents' desire to hunt, hike, and explore these lands. The city has expanded hiking and boating opportunities and loosened permit restrictions for using its watershed property. Although conflicts persist over access to particular parcels and policies, New York City has made significant strides in increasing recreational access to its watershed holdings. The change in recreational policy bespeaks a more fundamental shift. Long accustomed to viewing the Catskills as a sparsely populated region immune to larger economic and societal shifts—municipal officials believed that nature would protect water quality as long as people stayed away—New York City gradually adopted a more pragmatic stance that recognized the need to collaborate closely with watershed residents and upgrade technology to ensure the integrity of its water supply.

Water experts describe this more holistic approach as taking the "soft path" to managing water. By partnering with rural residents to protect water supplies and taking aggressive steps to curtail consumption, New York abandoned its long-standing practices of continuous supply expansion and reliance on large-scale technologies to ensure the delivery of high-quality water. The "hard path" to water security that New York City constructed remained in place—it continued to upgrade aqueducts, reservoirs, and treatment systems to meet the needs of its citizens. But protecting water at its source became the centerpiece of the city's approach to managing its sprawling supply network.[19]

The slow transition in environmental governance provides the arc of the narrative and also dictates the form and content of this book. A work of social, political, and environmental history, *Empire of Water* is an exploration of history from the bottom up and the top down. It analyzes developments from the perspective of those building and overseeing the water system and also from the point of view of those who lost their homes and businesses to satisfy New York City's demand for more water. An emphasis on political ecology inevitably highlights the actions of the engineers, lawyers, and politicians who had the power to remake landscapes. As a result, much of the first half of this book centers on the city's own decision-making processes and its legal battles with other states. New York's ability to secure state and federal backing for its waterworks projects was a critical component of its success. The path to Catskill water went through Albany and the United States Supreme Court. The latter chapters focus on management of the system and highlight the efforts of New York City and watershed residents to adjust to the new ecological expectations and financial circumstances that emerged in the 1970s.

Environmental historians are frequently accused of writing declension narratives in which they portray the natural world as idyllic and in balance before human beings came along and cut down trees, built dams, and generally threw nature out of whack. I do not seek to replace one overly simplistic description of environmental change with another. New York City's water system is a work in progress. The present state of environmental and political equilibrium will not persist forever. Some watershed residents resent New York City's incursions and view the current state of affairs as a violation of their rights. They continue to file lawsuits and closely monitor the city's water operations. Nonetheless, most would acknowledge that they benefit much more from the water system than they did only a few decades before. In an era of intense political friction, the success of the MOA offers hope that a brighter environmental and political future is within reach.

optimistic

CHAPTER ONE

From Croton to Catskill

O n the afternoon of July 15, 1890, Mayor Hugh Grant boarded his horse-drawn carriage en route to a ceremony marking the introduction of water from the New Croton Aqueduct into New York City's distribution system. Plans called for Grant "to appear as a fresh water Neptune" and turn a regulator, which would send a torrent of water into gatehouses at the main Central Park reservoir. City employees at an uptown gatehouse—not the mayor, whose planned whirl of the regulator was pure political symbolism—would initiate the flow of water into the reservoir. Faithfully obeying the instructions of Alphonse Fteley, chief engineer of the Aqueduct Commission, workers opened the gates of the new aqueduct just before two o'clock, ensuring that a gush of water would arrive at Central Park on the hour. When the mayor arrived a few minutes after two, he was disappointed to discover that a new era had begun without him.[1]

Grant's frustrating afternoon symbolized the difficulty of providing a pure and abundant supply of water to New Yorkers around the turn of the century. The root of the problem was the rapid growth and transformation of the city. The arrival of hundreds of thousands of eastern and southern European immigrants sharply increased demand for water. Lifestyle changes, particularly the growing popularity of apartment houses among middle- and upper-class residents, also strained the water system. Apartment houses provided residents access to more water-consuming devices, fueling higher per capita consumption. Each unit in the Stuyvesant Apartments, considered New York's first apartment building when it opened in 1870, included two water closets and a separate bathroom. Just a few blocks away, tenement dwellers shared hallway taps with several other families and relied on privies.[2]

The dynamism and energy of the city spilled over into the larger metropolitan area. As late as the 1830s, Brooklyn was a minor city, home to fewer than forty thousand residents. Six decades later, owing to an influx of residents from

New York and aggressive annexation of nearby communities, it had become one of the nation's largest cities.[3] Just as New York gradually expanded the Croton system, Brooklyn made piecemeal additions to its network of wells and infiltration galleries on western Long Island. These improvements fell far short of what was required to ensure adequate service. In 1895 the *Brooklyn Daily Eagle* charged that the city "has been so near a water famine within the last three years that men in power were afraid to let the public know the exact condition."[4] The lure of improved water supplies motivated many residents of Brooklyn and Staten Island to vote in favor of annexation by New York.[5] The 1898 union of Brooklyn, New York, Staten Island, and surrounding communities created Greater New York, a sprawling metropolis that greatly increased the physical dimensions of the city. The city's population nearly doubled to 3.5 million. Filling the tanks and tubs of its newest residents required New York to undertake a massive expansion of its water network.

Nature graced New York with several options for increasing its supplies. Municipal officials considered tapping aquifers on Long Island; the Hudson River; Connecticut's Housatonic River and its tributaries; streams in Dutchess County, north of the Croton region; and the rivers and creeks of the Catskill Mountains. Engineers weighed several criteria, including water quality, expected yield, and cost of delivery. But the differences among the sources were less significant than what they had in common, namely that local communities had no intention of ceding their current or future water sources to New York City.

Gaining access to new sources emerged as the most challenging task facing New York's water overseers. Both the private and public sectors stymied the city's bid to secure a new supply. The Ramapo Water Company purchased water rights from property owners across wide swaths of the Catskills, effectively preventing New York from tapping productive mountain streams for a public supply. State lawmakers supported new restrictions on New York's right to develop streams and aquifers lying closer to the city. The future of the water supply and the viability of New York City itself hinged on overcoming the reflexively antiurban bias of the state legislature.

Despite the delay in tapping new watersheds, the turn of the century was an unusually active period for the existing water system. The Aqueduct Commission oversaw construction of several new reservoirs in the Croton region. These reservoirs remade the northern portion of Westchester County and much of Putnam County, wiping out hundreds of acres of fertile farmland, displacing residents, and sundering communities. The effects of extending the water system were not limited to the Croton region. They also reverberated within the city, where officials had constructed a network of reservoirs and pumping stations to enhance water delivery. Upgrading the Croton system obviated the need for some of this infrastructure, allowing the city to redeploy it for other uses. The decision to dismantle an urban reservoir marked the beginning of the recycling of water infrastructure, a land-use practice that transformed New York and its suburbs throughout the twentieth century.

The extension of the Croton system and the drive to secure additional watersheds reconfigured the physical and political landscape of southeastern New York State. By 1905, when the city's campaign for a new water supply shifted into high gear, the Aqueduct Commission was nearing completion of the sprawling New Croton Reservoir, the centerpiece of the expanded Croton network. New York added two more reservoirs over the next several years, finally completing the water system it had begun constructing in 1837. The build-out of the Croton system marked the end of an era. A new set of political arrangements emerged in which the interests of the budding suburbs began to overlap with those of New York City. This urban-suburban alliance spurred the development of an entirely new water system in the Catskill Mountains. The construction of this new water network transformed distant rural watersheds, catalyzed suburban growth, and remade the built environment of the city.

Croton, Ramapo, and the Politics of Water Supply

The completion of the New Croton Aqueduct allowed more water to be conveyed under higher pressure, raising hopes that the Aqueduct Commission had delivered on its promise of a reliable water supply. These hopes proved short-lived. New Yorkers took full advantage of the bounty, leading to unprecedented increases in overall and per capita consumption. The increased delivery capacity of the new aqueduct proved a boon when water was abundant, but as the *New York Times* trenchantly observed, "A large aqueduct and a liberal flow only increase the danger of periodical exhaustions of the supply so long as there is inadequate provision for the storage of the surplus of wet periods."[6] Construction of new reservoirs that could store large amounts of water lagged behind completion of the aqueduct. The commission had, in effect, provided New Yorkers with a much more powerful straw but only marginally increased the volume of water in the city's glass. Ironically, the city experienced a severe water shortage in the fall of 1891, only a year after the new aqueduct started delivering water. The commission spent the next two decades playing catch-up, damming the Croton and its tributaries to build nine additional reservoirs.[7]

The development of the Croton system submerged large sections of Westchester and Putnam Counties. The scale of system expansion far surpassed any construction project that the city had ever undertaken. Nevertheless, by the mid-1890s, both the public and private sectors had turned their attention to the next water source. Increases in water consumption continued to outpace the construction of new reservoirs. To sustain its breakneck growth, New York needed to look beyond the Croton for more water.

Even before it began to seriously investigate new sources, the city found itself outflanked by rural communities wary of becoming New York's next watershed. In 1896, the state legislature passed the Burr Act, which barred Brooklyn from tapping the waters of Suffolk County, in eastern Long Island. The bill compounded Brooklyn's acute water difficulties. Brooklyn residents hoped that political union with New York would finally resolve their long-standing water problems.

Private interests also anticipated the increased demand for water in the metropolitan area. Whereas Suffolk County blocked Brooklyn's water expansion to protect the oyster beds and tourism that were the mainstays of its own economy, the Ramapo Water Company sought to cash in on the city's thirst. Ramapo's attempt to monopolize water sources vital to New York's future featured all the classic elements of Progressive Era melodrama: avaricious private interests; an aggressive press corps; a maverick public figure in City Comptroller Bird Coler; and a valiant civic organization, the Merchants' Association of New York. The battle over Ramapo's plan to tap mountain water sources launched the movement to secure a new water supply for the city.

In 1895, the state legislature granted the company extraordinary powers to acquire water rights throughout New York and to contract with municipalities to sell the water it collected. Ramapo purchased water rights from landholders throughout southeastern New York State, particularly in the Catskill Mountain region, which had been identified as early as 1886 as a potential water supply for the city.[8] By the late 1890s, the company had secured the rights to enough water to contemplate the construction of a delivery system capable of providing New York City with two hundred million gallons of water a day (MGD), nearly half the total consumption of the five boroughs.[9] With insurers and hoteliers urging Mayor Robert Van Wyck to significantly expand the water supply to reduce the likelihood of fire, in 1899 Ramapo asked the Board of Public Improvements, comprising the leaders of municipal departments responsible for public works, to sign a contract with the company to supply New York with water from the Catskills.[10] The company had the ear of Tammany Hall, the notoriously corrupt Democratic Party machine that had regained control of municipal government with Mayor Van Wyck's victory in 1898. With the explicit backing of state law and the support of Water Commissioner William Dalton, the company appeared poised to transform New York's water system from a publicly run utility to a public-private operation in which the city conveyed water from the Croton watershed while Ramapo delivered it from the Catskill Mountains.

The company greatly underestimated the extent and intensity of the opposition that its plan would arouse. In August 1899, Comptroller Coler, a member of the Board of Public Improvements who harbored deep suspicions about Ramapo's intentions, succeeded in postponing a vote on the proposed contract. After a two-week delay, Coler presented the conclusions of a series of engineers' reports that he had hastily commissioned. Coler sought to debunk the claim that New York was on the verge of outstripping its water supply. One report concluded that New York could take its time in choosing its next water source: "There is, therefore, no question of sufficiency of supply in 1904, or 1909 . . . and, consequently, there is no need of excessive haste in this important matter."[11] Other engineers questioned Ramapo's most basic claims, arguing that the project would yield substantially less than the 200 MGD promised, and that the city could build the project itself for approximately one-fifth the expense.[12]

Considering the circumstances under which the reports were produced—at the behest of an opponent of the Ramapo plan and under an absurdly compressed two-week deadline—their conclusions were more educated guesswork than well-supported assertions. Nonetheless, they set the terms of the debate on what the *New York World* dubbed "the Great Chartered Ramapo Robbery."[13] The newspaper took the unusual step of formally entering the political arena by securing an injunction prohibiting the city from signing a contract with Ramapo.[14] Then, with the injunction deadline looming, the Merchants' Association, a membership organization composed of influential businessmen with an interest in public policy, persuaded the Board of Public Improvements to table consideration of the contract for several months, pending completion of a comprehensive analysis of New York's water supply.

The Merchants' Association report, whose great length—627 pages—seemed to attest to its comprehensiveness and veracity, proved the death knell for the Ramapo plan. Newspapers trumpeted the report's claim that the proposal would cost the city $195 million more than a publicly built system. One headline, "Ramapo Raid Would Rob New York City of $195,460,070," suggested that the report told the press what it wanted to hear, and that its conclusions were accurate to the dollar.[15] A more skeptical observer might have noted the similarity between the estimate and the figure cited in a Merchants' Association fund-raising solicitation issued immediately before it undertook its investigation. In its appeal, the association requested funds "for effective protection of the public interests against a possible needless burden of $200,000,000."[16]

These arguments proved especially effective because they confirmed the prevailing Progressive Era presumption that what benefited private interests often conflicted with the public good.[17] Ramapo opponents convinced the public that nefarious private interests sought to "procure a vast and profligate contract with the

City of New York through the connivance of city officials and the abuse by them of their official power."[18] Riding a wave of public outrage, anti-Ramapo interests sought to revoke the company's charter. In a letter to Republican gubernatorial candidate Benjamin Odell, Merchants' Association president William King highlighted both the 1895 legislation that conferred extraordinary powers on Ramapo, and language in the city charter that effectively deprived New York of the right to acquire additional water sources by eminent domain to expand its supply network. Without these powers of condemnation, King warned, New York "may be compelled to resort to a contract as the only means of procuring an increased water supply."[19]

The Ramapo Water Company fought vigorously to maintain its charter, but after the release of the Merchants' Association report in August 1900 and its enthusiastic reception by the New York press, most legislators regarded the company as politically toxic. In a brief distributed to lawmakers, the company refuted the claim that it enjoyed special powers not available to other corporations, arguing that "the proposed repealing measure is entirely without argument or reason for its support."[20] The company misread the prevailing political winds: the handful of legislators who may have read its plea likely viewed the highly technical comparison of the company's charter with existing state law on corporations as entirely irrelevant. At its core, the Ramapo controversy stemmed from the widespread perception that the company sought to exploit its exclusive control of potential water sources to egregiously overcharge New York City for access to these waters. Ramapo's focus on legal details amounted to an exercise in irrelevance. In the spring of 1901 the New York State Senate voted to repeal the company's charter by an overwhelming vote of forty-two to four.[21]

Despite arousing what might be described as justifiable hysteria in regard to New York's future water supply, the Ramapo episode left the most essential questions unanswered. Where would New York obtain its new water sources, and under what restrictions? These questions were not likely to be answered while Mayor Van Wyck remained in office. In 1902, New Yorkers elected Seth Low, former mayor of Brooklyn and a noted reformer. Shortly after taking office, Low commissioned John Freeman and two other prominent engineers, William Burr and Rudolph Hering, to prepare a report on New York's water system.

As the engineers embarked on their study of New York's long-term water prospects, the city worked to shore up its existing supply and delivery system. The New Croton Reservoir, a massive project begun in 1892, was finally nearing completion. Directly upstream, work on the Muscoot Reservoir was under way. And on Long Island, laborers retooled Brooklyn's aging water system. In 1899, the city began repair work on the leaky Millburn Reservoir, the most glaring symbol of Brooklyn's flawed water network.[22] New York built a mechanical filter plant to purify supplies from two Long Island ponds that had become contaminated, resulting in the de-livery of "seven millions of gallons of water filtered from localities which were previously a serious menace to the community."[23] Nature also cooperated, grac-ing the metropolitan area with several consecutive years of above-average rainfall. But these were only stopgap solutions. Within a few years, the Croton system would be fully developed, and the generous rains would inevitably fade. New York needed more water, and it needed it soon.

Reports, Responses, and Results

Voluminous and highly technical in parts, the engineers' report nevertheless reflected political realities. In a report he drafted for Comptroller Coler during the Ramapo controversy in 1900, John Freeman had identified the Ten Mile–Housatonic watershed, much of which lay in Connecticut, as the most promis-ing water source for the city, insisting that New York could overcome the legal obstacles associated with diverting water sources that originated outside state borders.[24] By 1903, legal precedent had established the practical impossibility of tapping interstate waters. The water commissioner therefore instructed Freeman and his fellow engineers (known as the Burr Commission) not to consider them as potential sources.[25] Politics, as much as hydrology, would determine the shape of New York's future water system.

The centerpiece of the report was its recommendation to develop two new wa-tersheds, one east of the Hudson River in Dutchess County, and the second in the Catskill Mountains. The engineers, well aware of New York's pressing need for additional supplies, prioritized construction of a waterworks network in Dutchess County, north of the Croton reservoirs. This new system would connect with the Croton supply, and eventually with the more distant Catskill system as well. Al-though the dense network of mountain-fed streams in the Catskills would provide much more water than the creeks of Dutchess County, the engineers favored tap-ping Wappinger Creek and other streams east of the Hudson first. The short-term appeal of this plan proved irresistible; New York could complete a waterworks system drawing on Dutchess waters within three years, thereby greatly reducing the risk of a water famine. In the battle against the Ramapo Company, engineers downplayed the risk of water shortages. With Ramapo dispatched, the urgent need for water reemerged as a paramount concern.[26]

The most significant news in the report, however, concerned the Catskill sourc-es, not the vastly inferior Dutchess ones. In 1900, Freeman had denigrated the Catskill waters that Ramapo intended to tap, observing of Esopus Creek, "I was myself very unfavorably impressed with the apparent lack of opportunity for large storage."[27] Two factors likely accounted for this negative appraisal. First, in the eyes of most New Yorkers, the Catskill sources suffered from simple guilt by association; anything Ramapo coveted was, by definition, suspect. Second, Freeman's original report reflected the knowledge gained from a rather cursory tour of the Catskill watershed, not the detailed examination that served as the basis of the 1903 report. Two letters from George Tauber, an engineer who investigated Catskill sources for Freeman's 1900 report, reveal Freeman's unfamiliarity with the region. In early November 1899, Tauber wrote Freeman asking him to clarify the exact locations on the Esopus where water samples should be drawn, observing that, although his notes called for taking water from Clove Bridge, the only similarly named place was Olive Bridge. Two days later, more familiar with the region, Tauber clarified the local nomenclature: "the *Post Office* is *Olive Bridge* but the *depot* is *Olive Branch*."[28]

In 1903, Freeman identified a number of factors that made the Catskill watershed an ideal source for New York's water supply. Tasteless, odorless, and colorless, the water was perfect. So was the landscape. The combination of high elevations, steep slopes, a heavily forested watershed, and rocky soil ensured that an exceptionally high percentage of precipitation found its way into the streams and creeks. The engineers, whose careful report teemed with caveats and qualifications, did not hedge when it came to the Esopus Creek: "The yield of this stream is phenomenally high."[29]

The enthusiastic endorsement of Catskill sources underscored the inherently political nature of resource evaluation. Ramapo no longer represented a threat, and the number of potential water sources had narrowed significantly. Catskill waters were no less pristine in 1900, but with the options for substantial new water sources narrowing to the Hudson River region and Catskill sources, the purity of mountain streams assumed center stage in the commission's report. Changing political dynamics recast the ecology and geology of Catskill streams.

In making the case for reaching ever farther for additional supplies, the engineers had to explain why the most obvious source, the Hudson River, was not a suitable alternative. Beginning in the 1870s, some New Yorkers had begun to promote the Hudson River as the city's best option for a new supply.[30] It flowed right through the city and contained an enormous volume of water, more than enough to sustain even New York's skyrocketing rates of consumption. But the Hudson had some significant drawbacks. Municipal officials, the business community, and the public demanded a high-altitude supply that would obviate the need for widespread use of private pumps to create sufficient water pressure. Hudson waters would have to be pumped to a higher altitude before delivery to the city. The city would have to actively manage these pumps, and its record did not inspire confidence. The second concern was the penetration of saltwater up the Hudson.

Because the lower Hudson was "really an elongated arm of the sea," the intake point for the water system would have to be located far enough upstream to ensure that saltwater did not contaminate the supply.[31] New York would still run the risk of saltwater penetration in late summer when the flow of fresh water diminished substantially, allowing seawater to migrate farther upstream.

Any West Side resident could cite the most serious objection to tapping the Hudson: the river was horribly polluted. Sewage, industrial contaminants, offal, and other pollutants combined to produce "jet-black masses of erupting gases" that emanated from the Hudson.[32] With an eye to both the near and distant future, the commission adopted a nuanced attitude with regard to the river. Recognizing that New York would very likely have to draw on the Hudson at some point in the future, the commissioners offered a lukewarm endorsement of its potential: "the use of filtered Hudson water is a practical possibility."[33] The cities of Albany and Poughkeepsie relied on the Hudson for their water supplies, and a significant reduction in the death rate from typhoid fever in Albany suggested that New York could tap the Hudson without jeopardizing the health of its residents.[34] Despite their productivity, Catskill Mountain sources would slake the growing city's thirst for only a generation, requiring New York to seek more water in the 1930s. With the Delaware River presumably off limits because of interstate legal concerns, only the Hudson would remain a viable alternative.

Nevertheless, Freeman and his colleagues rejected the Hudson as a short-term answer to New York's water woes. Acknowledging that even after factoring in the cost of constructing and maintaining filtration and pumping systems the river was likely a cheaper source of water than the Catskills, they argued that the vastly superior quality of Catskill water eliminated the Hudson from consideration. Using dirty Hudson water as a foil, the report fetishized the physical characteristics of Catskill water. In a section focused on dam construction, not water quality, the engineers noted the "remarkably soft water of the Esopus."[35] The engineer who led the Hudson investigations suggested that the decision not to perform a detailed technical and financial analysis of filtering water from the river reflected direction from his superiors, not a lack of technical resources: "The equipment of the department was such, however, that these questions could have received proper attention had occasion required."[36] Just as Freeman failed to fully explore Catskill sources in his 1900 report, he and his fellow commissioners—aware of the public resistance to tapping the polluted Hudson as a water source and enamored of the potential of the Esopus and other Catskill streams—elected not to conduct a detailed investigation of the river.

Thus, by the end of 1903, when the Burr Commission completed its report, a combination of legal, scientific, and political considerations had narrowed New York's potential water sources to just two: the streams of Dutchess County and the creeks of the Catskills. The contents of the report and the reputation of its authors ensured that it dictated the terms of the water debate. After the Ramapo controversy, New York sought to base its case on science and prevailing legal principles. With the release of the report, the seesaw had tipped back firmly in the direction of politics: by what means would the city obtain the authorization to develop the water sources identified in the report?

The challenge fell to Mayor George B. McClellan Jr., son of the famous Civil War general and a Democrat with close ties to Tammany Hall. McClellan's name is forever tied to the city's water supply because the bill that authorized the Catskill system was known as the McClellan Bill, or the mayor's bill.[37] This proved an ironic designation, because the bill that ultimately passed the legislature differed quite substantially from the one originally drafted by the mayor. Changes in the legislation reflected a basic political reality: the rural lawmakers who controlled the state legislature would determine the conditions under which New York would develop a new water supply, not McClellan or other municipal officials.

The recommendation to tap distant rural water sources pitted downstate against upstate. Although the specifics differed, the story line was unchanged: tapping country streams for urban water supplies threatened to destroy the rural economy. After publication of the Burr Commission report, Dutchess County legislators and business interests anticipated that New York City would act quickly on the recommendation to dam Fishkill and Wappinger creeks as the first stage in its water expansion project. From the city's perspective, these creeks were the closest viable water source; to the residents, factory owners, and officials of Dutchess County, they represented a critical economic lifeline. County political leaders appealed to the legislature for help.[38] Foreshadowing arguments it would later deploy in the Catskills, New York City claimed that its project would actually benefit manufacturing by building storage reservoirs that would better regulate stream flow. Rural lawmakers were not swayed. The city's mistreatment of Croton residents and their property was legendary. Legislators overwhelmingly approved the Smith Act, prohibiting New York from tapping Dutchess County streams. John Freeman could only lament that the city's excesses had scotched his second proposal for expanding the water system: "I suspect that the policy of the City has been in the past almost brutal in the force with which it has applied its time and its right under law."[39]

As 1905 dawned, New York City faced a daunting prospect: the Smith Act and the Burr Act prevented it from tapping all in-state watersheds within one hundred miles of the city, likely consigning it to a water famine when plentiful rains slackened. Catskill lawmakers planned to file a similar bill. If successful, it would deprive New York of all the water sources it had contemplated developing. The only option left would be the filthy waters of the Hudson. At the most basic level, the conflict reflected representative democracy in its most elemental form: legislators acting to protect the interests of their constituents. As Robert Grier Monroe, the city's former water commissioner, ruefully observed, "The City is so weakly represented in the Legislature that the countrymen can get anything through they want to."[40] Although the city's boorish behavior in the Croton watershed generated considerable skepticism about its promise to deal considerately with residents and businesses affected by future waterworks construction, opposition seemed to stem from a more visceral sense that New York's needs and those of rural residents were fundamentally incompatible. This issue of compatibility and the balance between urban growth and rural preservation lies at the heart of the most significant episodes in the development of New York City's water network in the Catskill Mountains.

To the Catskills

In contrast with the Croton watershed, where municipal water engineers and employees waged an energetic campaign in the 1890s to remediate pollution linked to increasing development, the Catskills seemed like a place from a bygone era. Even as it gained popularity as a summer resort destination for city residents, it remained a sleepy place for much of the year, its small year-round population consigned to eking out a living on marginal farmlands or working in small manufacturing plants. The hide-tanning boom that had swept the region in the mid-nineteenth century had long since petered out, leaving behind a transformed environment.[41]

Well before New York City officials eyed its waters, another natural resource of the region—its abundant stands of hemlock—had attracted the interest of outsiders. Beginning around 1816 and continuing into the 1870s, tanning operators stripped wide swaths of the Catskills of their hemlock trees for the tannin contained in their bark. They then applied the tannin to animal hides purchased from South America, churning out a variety of leather products for the booming American economy. In addition to its profusion of hemlocks, the Catskills contained many clear-running streams and plenty of limestone, two key ingredients in the tanning process.[42]

The tanning boom set in motion what Catskill historian Alf Evers called "a bombardment directed at the soil, the air, and the water of the mountains."[43] Deforestation promoted soil erosion and raised stream temperatures, leading to the disappearance of species such as trout that have low tolerance for warmer waters. Fish species that did survive struggled to prosper in streams and creeks laden with tannery wastes. Those trees not felled by loggers often succumbed to the fires that ravaged a region desiccated by deforestation.[44] Many manufacturers who had relied on Catskill creeks to power their plants switched to steam power after widespread cutting of hemlocks reduced stream flow. Charles Carpenter, hired by the state in 1886 to study the region's forests, described the environmental devastation left in the tanning boom's wake: "After the bark peelers passed through, millions of the best hemlock timber lay rotting in the woods, and millions more feet were consumed by the terrific fires which swept through in the way prepared for them."[45] One factory owner in the Ulster County town of Wawarsing attributed the reduction in water flow to the cutting of trees on swampy land, which dried out after being cut over. Carpenter noted that many businessmen shared the manufacturer's plight: "Examples could be multiplied all through and around this region."[46]

Tanning was only one contributor to the widespread assault on Catskill forests. Hundreds of sawmills churned out lumber for construction and to supply the many wood-using industries based in the Catskills. Factories turned the forests into baseball bats, furniture, barrels, railroad ties, and a host of other products. The diversity of the manufacturing base created a market for both hardwoods and softwoods, leading loggers to cut white pine, red spruce, sugar maple, black cherry, northern red oak, and other species to satisfy industrial demand. Cutting for charcoal and cordwood put additional pressure on Catskill forests. By the 1880s, only a sliver of original growth remained; the forest consisted primarily of saplings and young trees too small to cut. The timber boom was over.[47]

Despite widespread deforestation, state officials recognized the region's potential as a future water source for New York and other cities. The 1885 legislation that created the Adirondack and Catskill Forest Preserves cited watershed protection as a principal reason for their establishment. The following year, *Scientific American* published an article outlining the basic plan that the city would eventually adopt in constructing its Catskill waterworks.[48] By the time Burr Commission engineer Walter Sears visited the region in 1903, the Catskills were once again heavily forested: "The forest, which 30 to 50 years ago was stripped from nearly all the mountains, has, to-day, again covered with a good growth from 75 to 80 per cent of the whole area."[49] State government accelerated this transition by purchasing large tracts of land whose owners had defaulted on their taxes and placing these parcels in the forest preserve, off-limits to logging and most forms of development. But state acquisition of Catskill forests was small scale and fairly haphazard; protecting water quality was not a major consideration. Nature played the leading role in transforming the Catskills. The abundant snowmelt and rainfall that made the area such an attractive watershed ensured the rapid reforestation of the Catskills. The region's landscape recovered at precisely the same time that New York City sought additional water supplies.[50]

Burr Commission staffers appreciated the Catskills' excellent water supply potential: strong stream flow, a heavily forested watershed, and low population densities made it an ideal water source for a booming city. In completing their surveys of the Catskills, they depended on the assistance of the region's residents, many of whom would come to strongly resent the city's waterworks development. To identify promising dam sites, the commission collected rainfall data at six locations in the Esopus watershed. Limited staff and the vastness of the watershed forced field engineers often to rely on locals, such as the postmaster at the Grant Hotel station, to take daily rainfall readings.[51] When the Ulster and Delaware Railroad allowed the commission to attach a rain gauge to one of its stations, it likely envisioned an increase in freight traffic from dam and reservoir construction.[52] Two years later, once it emerged that waterworks construction would require the relocation of a long stretch of Ulster and Delaware tracks, the company became New York City's most formidable adversary in legislative hearings.

The willingness of Ulster County residents to assist the commission stemmed from a mixture of simple politeness and self-interest. Landowners at Bishop Falls, which engineer Walter Sears identified as a potential dam site, agreed to permit the commission to dig on their property if workmen refilled the holes once investigations were complete. Recognizing his interloper status, Sears astutely hired local resident Eli Humphrey to broker assistance from Catskill residents: "Mr. Humphrey's acquaintance with localities, property lines and owners makes him a desirable assistant in this particular work."[53] Some locals farmed marginal parcels that yielded a meager income and welcomed the possibility that New York City would purchase their land. With the end of the boom in natural resource extraction, the Catskill landscape offered little hope of economic prosperity. One resident of Kingston, Ulster County's largest city and center of regional commerce, claimed that his rural neighbors "merely exist on their farms. . . . Now they have nothing, even the stone is quarried out. It would be a deliverance to them to sell out."[54]

Catskill residents generally viewed the prospect of waterworks from an economic perspective. While many feared the project would devastate them financially, others sensed an opportunity to profit from the construction bonanza. The diversity of opinion regarding the financial effects of massive dam and reservoir construction resulted in a weakened and belated campaign to thwart New York City's designs in the region. Defeated but not dejected by the passage of the Smith Act in 1904, city officials introduced a new water bill when the new legislature convened in January 1905. Even as the McClellan Bill moved through the legislature, Catskill residents failed to mount a united front in opposition. At a mid-February meeting of the Kingston Board of Trade (essentially a chamber of commerce), members rejected a resolution instructing their state legislators to introduce a bill to block New York City's plans. Opponents of the resolution argued that water-works construction would actually increase the amount of business flowing to Kingston. Although the Board of Trade ultimately endorsed the resolution, many members continued to support the construction project.[55]

Disagreements among Kingston's financial elite reflected differing perceptions of the region's financial geography. Opponents of the waterworks project argued that Kingston merchants would suffer when their principal customer base—the county's farmers—was displaced.[56] One local merchant, irked by the willingness of many of his fellow businessmen to support New York's plans, urged them to examine the flow of commerce more carefully, noting that most of the farmers who traded in Kingston hailed from precisely those towns where the dams and reservoirs would be constructed. Farmers from other parts of Ulster, he observed, purchased their supplies not in Kingston, which lay at the extreme eastern end of the county, but in other rural commercial centers. To assume that additional commerce linked to waterworks construction would replace the steady stream of revenue from the county's farmers was, according to one resident, to trade "a sub-stance for a shadow."[57]

Proponents of the water project interpreted the economics of the region quite differently. In addition to emphasizing the potential benefits that property con-demnation offered downtrodden farmers, they argued that waterworks construc-tion would stimulate the summer resort business that had become a cornerstone of the local economy. Former Kingston city engineer George Bell anticipated that boardinghouses swallowed up by the construction would relocate closer to railroad lines leading to Kingston, resulting in more business for city merchants.[58] Bell also predicted that the centerpiece of New York City's construction plans—the mas-sive Ashokan Reservoir—"will be a great attraction" to the summer residents who increasingly represented the Catskills' economic lifeline.[59]

Albany, Westchester, and a New Watershed for the City

Despite the divisions among Catskill residents, Ulster County's representatives in Albany filed legislation to bar New York from tapping the county's waters. The political landscape had changed, however, since Dutchess County had stymied New York's water expansion plans. Although he hailed from upstate, the state's new governor, Frank Higgins, did not share the antiurban attitude displayed by many of his predecessors. Worried that legislators were intent on choking off ur-ban growth, he staked out a middle ground between promoting New York City's expansionist dreams and protecting rural residents who would bear the brunt of waterworks development. Higgins envisioned that the state, through a new water supply commission, would act "as a sort of umpire between the two interests involved."[60] Under a bill promoted by Higgins, any city seeking to draw water supplies outside of its municipal borders would be required to obtain authorization of the state commission, which would allocate water between local communities and the city.

Higgins's desire for a state commission reflected the tendency of Progressive Era reformers to establish new boards and agencies to oversee powerful interests. In most cases, the intent was to enhance state supervision of utilities and other economic monopolies. But reformers also targeted government itself; inefficient and corrupt municipal governments could not be expected to provide vigorous oversight of the private sector. Federal, state, and city governments created new commissions and boards, often stocked with professors and other experts, to accomplish a wide variety of goals. Well aware of the mistreatment of Croton watershed residents, Higgins sought to use the new tools of reform to avert a repeat performance in the Catskills.[61]

The governor and the mayor vehemently disagreed about the path forward. Although he embraced elements of the reform agenda, Mayor McClellan vigorously objected to the commission, arguing that it would lead to significant delays in developing new water sources. Higgins countered that the lack of a commission that could compel a municipality to pay indirect damages when it condemned land for water supply purposes was an insuperable obstacle to securing the consent of watershed communities: "With the taking and destruction of the factory on the water shed the store may lose its customers, the church its worshippers, and the village its population. Values may thus be depreciated without compensation where there is no actual taking of property by the City."[62]

With Higgins and McClellan deadlocked on the state commission, others rushed in to forge a compromise. Assemblyman and chair of the Water Supply Committee George Agnew found himself pulled between two masters. As a West Side resident he sought to represent the city's interests, but as a Republican legislator he was charged with drafting legislation that would garner the support of his political comrades from the state's rural areas. In January 1905 he introduced a bill that would establish a temporary commission to study the wisdom of establishing a permanent state water supply commission. After Governor Higgins made it clear that a new water supply for New York City hinged on the establishment of a commission with significant powers, including the right to mandate payment of indirect damages, Agnew drafted a more ambitious and detailed bill. Before revising his bill, he had to convince city officials that a state water supply commission would serve their interests. In a draft of a letter to municipal officials, Agnew observed of the proposed commission, "I believe it will tend to tranquilize the up State property interests, and thus bring us nearer the desired goal of an adequate water supply, which will be ample in years of low rain as well as in years of plenty."[63]

City officials publicly opposed the state water supply commission, but privately they worked closely with Agnew to tailor the bill to their liking. A marked-up version of the bill in Agnew's personal papers includes numerous edits that he noted were "city and aqueduct suggestions," changes requested by municipal officials and members of the Aqueduct Commission. These modifications were aimed at reducing damage payments and expediting the state approval process. One critical new provision required the state commission to rule on applications for a new water supply within ninety days.[64] The bills creating the state commission and granting New York the right to tap Catskill sources were paired and numbered Chapter 723 and Chapter 724, respectively. Without the intervention of Higgins and Agnew, who receive virtually no mention in most accounts of the legislation, it is highly unlikely that the legislature would have approved the McClellan Act.[65]

By creating a mechanism for state control over water resources, Higgins and Agnew began to change the terms of the debate in Albany. Granting New York the right to tap the Esopus Creek and other waters in the Catskills no longer meant placing Ulster County at the not-so-tender mercies of the city. Once the idea of a state commission began to gain some traction in the legislature, Higgins began to lose patience with what he considered Ulster County's exaggerated claims, observing in late March, "I have found that they fear conditions which will not exist. Misleading statements have been made about the danger to Ulster County from the encroachment of the New-York water supply system."[66]

A state commission alone could not induce rural legislators to drop their resistance to New York City's water development plans. After Mayor McClellan visited Albany in late February 1905 and all but insisted that the legislature approve his bill as written, passage of the legislation appeared unlikely. In response to McClellan's reluctance to substantially modify his water bill, Assemblyman J. T. Smith, sponsor of the previous year's bill to prevent the city from tapping Dutchess County sources, remarked, "I guess Ulster and Dutchess and the other counties will have to stand together."[67]

Behind the scenes, however, self-interest took precedence over rural unity. Westchester County had embarked on the path that would make it one of the country's archetypal suburban enclaves. The cities and villages of the county's southern tier were booming. In New Rochelle, on the Bronx border, contractors grumbled about the lack of qualified workers. The city's newspaper heartily endorsed the economic upsurge: "Land which has been idle for many years and not even used for farming purposes is being rapidly developed into magnificent residence sections in all parts of the city."[68] Lured by improved rail connections to Manhattan and the availability of land, New Yorkers flocked to Westchester, whose population increased by more than 50 percent from 1900 to 1910.[69]

The influx of residents strained Westchester's water supply network, which was owned and managed mostly by private companies. With the Croton River off-limits, these companies relied on sources close to population centers. Rapid residential development began to encroach on these watersheds, threatening the purity of these sources. Faced with a choice between stunting economic growth and risking their health, Westchester residents devised a clever alternative: tap into New York City's water system. Connecting to city aqueducts would provide Westchester communities with an abundant and pure source of water. It would, in essence, return the Croton River to its rightful owners. Moreover, it would eliminate the need to set aside additional land for water supply purposes and unlock the watersheds of the private water companies for future development. The *Westchester News* clearly articulated the emerging suburban mind-set that viewed forests and pastures as real estate waiting to happen: "The land in the lower part of the county is too valuable to be taken as a watershed."[70] In February 1905, Assemblyman Wainwright filed a bill in Albany giving Westchester communities the right to connect to New York City's water system at reasonable rates.

The farmers who predominated in the northern reaches of the county, where the Croton reservoirs were located, had little interest in tapping into the city's water network. They wanted to fight New York, not cooperate with it. New York had acquired large portions of northern Westchester for its reservoirs and buffer lands; its refusal to pay property taxes on its holdings, a clear legal obligation, imperiled the delivery of basic services in these communities. Residents also chafed at New York's abusive treatment of farmers and their property, particularly the practice of burning barns and endangering livestock on property recently acquired by the city. The *Mount Kisco Recorder* distilled the bitterness of area residents: "It is hightime [*sic*] some move was made to bring the tax-dodging city of New York and its barn-burning, lying and swindling employees to terms."[71]

Westchester's representatives in Albany appeared to be listening. Assemblyman James Apgar, who had previously shown little inclination to resist the expansion of the Croton system, filed bills to prevent New York from building additional reservoirs in Westchester. Apgar argued that Westchester could no longer afford to let New York divert the county's best water sources, because Westchester needed the water for itself.[72] Representatives from Putnam County followed suit. If these bills passed, New York City would be prohibited from constructing the final two reservoirs in the Croton system. Not only might New York be prohibited from tapping Catskill waters, it could also lose access to Croton streams that had long figured in its expansion plans.

In early March 1905, New York City assistant corporation counsel Charles Guy, who remained in Albany throughout the legislative session to shepherd the water bill, convened legislators from the affected counties to resolve the impasse.[73] Guy recognized the impossibility of any overarching solution to rural resistance. He brokered a deal by granting specific concessions to affected counties. He agreed to permit Westchester and Putnam County municipalities to tap into New York's water supply. Communities could connect to the Croton system or tap into the main aqueduct that would run from the Catskills through Putnam and Westchester en route to the city. He also guaranteed that New York would refrain from building additional reservoirs in Putnam and Westchester Counties for sources located within these counties. He overcame Dutchess County's resistance by removing a clause from the McClellan Bill that would have effectively repealed the Smith Act and allowed the city to develop Dutchess water sources. Finally, he assured Ulster that New York would pay adequate damages. In exchange for these concessions, New York City would be permitted to construct the final reservoirs in the Croton system and, most important, develop a new water network in the Catskill Mountains.[74]

Just as the pact split the once-solid rural alliance in the state legislature, it also exposed fissures in Westchester County. The grand bargain effectively disenfranchised communities in northern Westchester, which had little need of the city's water and faced the prospect of yet another large reservoir in their midst.[75] But the deal advanced the interests of the growing communities in southern Westchester. Assured of a plentiful supply of water, the county could pursue its suburban destiny. Westchester politicians adopted a shrewd strategy. In arguing for restrictions on New York City's right to secure additional water sources, they played the role of aggrieved rural county. But by securing the right to tap into the city's expanded supply network, they acquired water sources essential to suburban growth. Meanwhile, Putnam and Dutchess legislators could return home to their constituents claiming to have finally slain the New York water dragon. Lawmakers from affected counties urged their legislative colleagues to support the compromise they had forged, explaining that the revised bill addressed their particular local interests, while the sister legislation creating the State Water Supply Commission guaranteed the future protection of rural areas from New York and other thirsty cities.

Not surprisingly, Ulster County, sacrificed on the altar of political self-interest, felt betrayed by its rural brethren. By late April, the *Kingston Daily Freeman* had adopted the bitter tone of a jilted lover: "Loud were the cries that up to this year came from Westchester and Putnam, but the finest and most exquisite sort of politics has been played by the Tammany administration of New York city [*sic*] in the furtherance of this measure. Those very elements which should be united with us in the opposition to this bill are now in favor of it and are opposed to us."[76] Despite obtaining additional concessions from New York City that severely curtailed the scale of waterworks construction in Ulster County, its legislators continued to oppose the bill. But the creation of the state commission and the support of the metropolitan area counties persuaded the legislature to pass the bills. With a new water source for New York finally within reach, politicians and legislation gave way to engineers and blueprints.

Reuse of Space and the Establishment of the New York Public Library

The imperative to increase the supply of water flowing to New York focused public attention on the construction of new dams, reservoirs, and aqueducts in rural areas. To most New Yorkers these watersheds were an abstraction; they would likely never see the Croton River or Esopus Creek. But as they moved about the city, urban dwellers encountered frequent reminders of the water supply.

Water infrastructure was a defining feature of the late nineteenth-century cityscape. The elegant High Bridge, completed in 1848, carried the Old Croton Aqueduct over the Harlem River into Manhattan. Central Park contained two reservoirs, and the commanding Ridgewood Reservoir, perched high on the border of Kings and Queens Counties, supplied water to Brooklyn's burgeoning population. The most famous reservoir and one of the city's most striking structures was the Croton Reservoir (also known as the Murray Hill Distributing Reservoir), whose four-story gray walls and two large basins loomed over Fifth Avenue and 42nd Street.

The Croton Reservoir embodied New York City's mid-nineteenth-century aspirations to cosmopolitanism. The chief engineer of the original Croton project, John Jervis, had collaborated with engineer and architect James Renwick Jr. to create an urban reservoir whose cornices, towering center pylons, and impressive scale evoked New York's connection to Egypt and other ancient hydraulic civilizations.[77] Chosen for its location at the highest point of Murray Hill to increase water pressure to densely populated downtown districts, the reservoir was an odd symbol of urban accomplishment. When completed in 1841, it had few neighbors and towered over the handful of scattered structures in the surrounding area. Across Fifth Avenue lay "an open field, upon which stood a single country house."[78] When Augustus Fay painted the reservoir and its surroundings in 1850, the immediate neighborhood still retained its rural character, but new brick buildings loomed in the distance, signs of the encroaching city.

Fig. 2. Augustus Fay, *View of the Croton Water Reservoir, New York City*, c. 1850. The reservoir was located at Forty-Second Street and Fifth Avenue; the surrounding area was still largely undeveloped in the mid-nineteenth century. (I. N. Phelps Stokes Collection, Miriam and Ira D. Wallach Division of Art, Prints and Photographs, the New York Public Library, Astor, Lenox and Tilden Foundation)

By the 1860s, this vital component of New York's infrastructure doubled as an elite gathering spot. The parapet that overlooked the basins provided an ideal aerie from which to survey New York's growth. Affluent New Yorkers promenaded on weekend afternoons, displaying their finery and taking in the increasingly crowded vista of the developing city. Novelist E. L. Doctorow memorably depicted the reservoir's heyday: "New Yorkers loved their reservoir. They strolled along the parapet arm in arm and were soothed in their spirits. If they wanted a breeze in summer, here is where it would blow. Puffs rippled the water. Children launched their toy sloops."[79]

The reservoir's popularity could not withstand the onslaught of urban growth. Fifth Avenue had become one of New York's most exclusive streets, rendering the reservoir an eyesore.[80] When the water department connected the reservoir's Murray Hill neighborhood to a new high-elevation water distribution system in 1876, it ruptured any remaining affection residents had for the reservoir.[81] No longer reliant on the reservoir for their water, property owners appealed to city officials for the removal of what the *Brooklyn Daily Eagle* would later dub "a relic of a former municipality."[82] This first concerted attempt to tear down the reservoir led those on both sides of the issue to craft arguments that framed the reservoir debate for the next two decades. Removal proponents claimed that waterworks improvements made the reservoir superfluous; the gray walls of the reservoir, they argued, should give way to a patch of green, thereby adding to the acreage of Bryant Park, which lay just to the west. Advocates of maintaining the reservoir countered that it remained an essential component of the water distribution system. In an 1878 speech, civic gadfly George Butler emphasized the need to retain the reservoir as a secondary water source in case an accident cut off supplies from the Croton watershed or "a convulsion of either nature or of society" disrupted the flow from the Central Park reservoirs.[83] The reservoir survived this first attack, but succumbed two decades later when some of New York's most powerful residents joined together to construct the city's flagship public library. Two forces—the reservoir's central location, and elite aspirations for the booming city—combined to launch the era of infrastructure recycling.

Various interests sought to claim the reservoir site for their own purposes, recognizing that as commercial and residential development in the city pushed ever northward, the once remote uptown location now represented prime real estate. The most notorious scheme called for covering the reservoir and placing a beer garden on its roof. In 1893 the state legislature voted to replace the reservoir with a library to be built with funds bequeathed by New York lawyer and former presidential candidate Samuel Tilden. Tilden hoped his gift would create a library to rival the one recently constructed in Boston and those of European capitals. The trustees of Tilden's will soon realized that the $2 million he designated for the library fell far short of the amount required to establish a first-class library worthy of the city, leaving the future of the reservoir site in doubt. In 1895, they joined forces with the city's most prominent research libraries, the Astor and Lenox libraries, to form the New York Public Library.[84] Both libraries boasted excellent collections, but neither had the space nor the resources to create a research library on par with those in Paris or London.

The decision to create a great public library for New York immediately raised an obvious question: where would this new library be located? In combining with the Astor and Lenox libraries, the Tilden trustees had forfeited their claim to the reservoir. After dismissing the Lenox Library as too cramped, the new library's board of trustees shifted its focus to the reservoir. In early 1896 the board approached Mayor William Strong to ask for his assistance in obtaining the reservoir site, and for a grant of $2.5 million to construct the library building.

Location was central to the public relations campaign for the new library. In their appeal to the mayor, the library trustees emphasized that the reservoir stood on "the most central and easily accessible spot on the Island."[85] Land that had been considered well uptown, on the fringe of the developed city when many of the trustees were children, now sat at the transportation crossroads of New York.

With Central Park occupying the center of the island from 59th Street to 110th Street, a location on either side of the park would prove inconvenient to a substantial number of New Yorkers. A real estate expert hired to find a central location for the new library reported that alternative sites would be prohibitively expensive. The library trustees argued that a gift of city-owned property would enable them to devote their limited resources to buying books rather than land.[86]

To urban boosters and the elite guardians of the Astor and Lenox collections, the need for a massive, amply endowed research library was glaring. Secure in its reputation as a center of commerce, New York's civic status now depended on its cultural achievements. Like civic leaders throughout the nation, the library trustees had fallen under the sway of the City Beautiful movement. They looked to shed the old gray image of the city symbolized by the reservoir and replace it with a library—a powerful symbol of New York's physical and cultural attainment. One magazine editor claimed that if City Hall withheld its support, New York would "remain, no one can say how many years longer, under the imputation of being the one great metropolis of Christendom which has no public library worthy of the name or adequate to its needs."[87]

Swayed by the trustees' arguments, Mayor William Strong and the state government endorsed the destruction of the reservoir. But the state legislation authorizing its removal required the consent of the Board of Aldermen, providing opponents one final opportunity to save the reservoir. The New York Board of Fire Underwriters, whose members insured valuable downtown offices and warehouses, led the fight to retain the reservoir. The underwriters read from George Butler's script, arguing that the reservoir allowed for rapid delivery of water to downtown districts in case of fire.

The controversy surrounding the reservoir thus focused not on the wisdom of the site as a suitable location for the city's flagship library, but about its utility as a water source. The underwriters insisted that removing the reservoir amounted to "an invitation to a conflagration."[88] By the 1890s, the Croton Reservoir conveyed only a small portion of the water distributed to downtown districts (most arrived directly via pipes from uptown), but firefighters could draw on its twenty million gallons to fight large blazes. At a public hearing and in written testimony, the Board of Fire Underwriters played on legislators' fears that removing the reservoir would greatly increase the risk of a catastrophic fire. Conveying their desperation typographically, the underwriters decried the idea that "THE SAFETY OF THE ENTIRE POPULATION SHOULD BE PUT IN JEOPARDY IN ORDER THAT A GRAND SCHEME OF BENEVOLENCE MAY BE CARRIED OUT."[89]

The conflicting perspectives also reflected a clash of aesthetic and cultural values. Supporters of maintaining the reservoir stressed that its value went beyond the utilitarian. Samuel McElroy, chief of the Brooklyn Aqueduct, informed the aldermen that "The structure itself is a most admirable specimen of Hydraulic architecture, and of the most substantial construction and as such ought to be preserved."[90] The voice of an older era, McElroy could imagine few structures more beautiful than the Croton Reservoir. Library supporters had little patience for this sentiment; New York's greatness depended on constructing new buildings that would enrich the cultural and educational life of the city, not preserving old ones of dubious architectural value. One library trustee expressed the new calculus of civic worth: "It is humiliating to think that the metropolis of America has in this combination of great libraries but one-ninth the number of books that the French institution has."[91]

The reservoir tussle also marked a clear transition in the visual hierarchy of public space. New, visually stimulating structures such as the Brooklyn Bridge, completed in 1883, clearly doubled as both infrastructure and art, and were widely recognized as tangible symbols of civic accomplishment. But municipal government increasingly shunted other components of urban infrastructure underground or to the outskirts of the city. In the early 1890s, after protracted legal battles, the city finally succeeded in forcing utility companies to bury all electrical wires. Streetcars and elevated rail would soon give way to subways.[92]

Water infrastructure reflected this subterranean trend. The original Croton Aqueduct entered the city over the High Bridge, a structure built expressly to carry water into the city; its successor, the New Croton Aqueduct, slinked into New York underground, ending at a comparatively modest gatehouse in upper Manhattan.[93] The expansion of the waterworks network in rural areas obviated the need for large-scale water storage within the city, making highly prized public space available for other uses. The aldermen recognized the educational and cultural benefits of having a world-class library, and approved the request to demolish the reservoir on the condition that the city install new mains to compensate for the reduced water flow. Once a monument to urban civilization, waterworks had become part of the invisible tangle of infrastructure that ran beneath New York's streets. Workers dismantled the reservoir, reusing much of the stone from its walls and basins for the library's foundation. Active construction began in 1902, and by 1911 New York possessed one of the world's premier libraries.

In the space of little more than half a century, New York City had changed almost beyond belief. The increasing glamour and sophistication of the city was one of the most notable developments. In 1849, perched atop the Croton Reservoir, poet Walt Whitman envisioned a New York "great in treasures of art and science . . . and in educational and charitable establishments." He also confidently predicted that the reservoir and the other components of the Croton system would last for "ages and ages."[94] These visions collided in the razing of the Croton Reservoir and its replacement with the New York Public Library. What neither

Fig. 3. New York Public Library, c. 1910. The flagship building of the New York Public Library was built on the site of the Croton Reservoir. It opened in 1911. (Library of Congress, Prints and Photographs Division, Detroit Publishing Company Collection)

Whitman nor most other observers could foresee was the tremendous growth of the city, growth that required continual expansion of the Croton system beginning in the 1860s. By the 1890s, the Croton Reservoir, which had once been the most prominent symbol of the water system, was a relic. One journalist declared that "a new era had come, when the antiquated reservoir of water must give place to the modern reservoir of books."[95] The reservoir may have been on its way out, but the city's need for water was only increasing. Municipal officials successfully redeveloped one small but significant parcel in the city. The question now was whether they could achieve the much taller task of developing distant rural watersheds to keep pace with the unrelenting growth of the city.

With the enactment of legislation authorizing the Catskill water system, New York joined other American cities, most notably San Francisco and Los Angeles, which were planning significant long-distance additions to their water systems. Los Angeles's development of the Owens Valley, fixed in the public mind by the film *Chinatown*, has become a byword for rapacious and underhanded exploitation by urban interests. Historians routinely cite the controversy surrounding

NY like SF + LA

San Francisco's decision to construct a dam in Yosemite Park for its water supply as embodying the tension between the impulse to preserve nature and the desire to harness it intelligently. The Raker Act of 1913, which gave the city the right to build Hetch Hetchy Reservoir in the park, is often seen as symbolizing the triumph of a decidedly utilitarian approach to nature.[96]

Although New York City also reached far into its hinterlands to sustain urban growth, its water supply trajectory resembled that of Boston more than West Coast cities. Both Boston and New York drew on increasingly more distant supplies of water, and in the process reshaped urban, suburban, and rural environments. In the 1830s, Boston and New York developed municipal water systems that tapped nearby rural water sources. The inability of these original water networks to satisfy escalating demand for water prompted both cities to undertake major infrastructure expansions in the twentieth century. In 1906, New York began work on its first Catskill dam. Three decades later, Boston embarked on the Quabbin Reservoir, the centerpiece of its expanded water system. In both cases, damming distant rural streams fueled urban growth and limited the need for waterworks development in suburban areas.[97]

The urban appropriation of rural resources gave rise to similar political and geographical shifts in both states. In Massachusetts, the need for water led to creation of a regional water authority for Boston and its suburbs. By granting Westchester County and other communities the right to connect to the city's water supply, New York became a de facto regional water supplier. The need for water accelerated the political realignments that would come to define twentieth-century politics in New York and Massachusetts, respectively: downstate vs. upstate, and east vs. west. Metropolitan areas tapped rural resources to ensure their continued growth. In conjunction with larger economic and technological changes, the drive to secure water supplies created a segmented landscape. The ability to transport products cheaply meant that water, milk, timber, and other commodities and natural resources increasingly came from remote rural areas. Improved rail connections and the desire for more comfortable living arrangements stimulated rapid suburban development in Boston and New York around the turn of the century.[98] The appropriation of rural water accelerated the reconfiguration of the suburban landscape. Tapping Catskill water had a paradoxical effect on Westchester and Putnam Counties. The ability to connect to New York City's water network facilitated rapid commercial and residential development, but it also preserved valuable open space that otherwise would have been needed for reservoirs to supply the expanding suburbs.

The segmentation of the landscape extended into the city itself. The decision to demolish the Croton Reservoir and replace it with the New York Public Library initiated a decades-long process of infrastructure erasure. Extending its reach into high-altitude rural areas allowed New York to dispense with many urban reservoirs and pumping stations. Waterworks expansion thus resulted in a double boon for New Yorkers. In addition to gaining access to high-quality rural supplies, they could patronize the libraries and parks that took the place of water system infrastructure.

If the exploitation of rural resources reshaped urban and suburban landscapes, the needs of the city, in turn, dictated how the countryside's assets were extracted and protected. Waterworks expansion in the Catskills differed quite markedly from the scattershot approach to development that prevailed in the nineteenth century. Public policies increasingly reflected the priorities of urban areas, priorities that on their face could seem paradoxical. But cities prized both development and conservation. The establishment of state and national parks in the final decades of the nineteenth century reflected the rise of an urban environmental agenda that emphasized the need to preserve particular places for the enjoyment of city dwellers. At times, as in the establishment of the Adirondack and Catskill Forest Preserves, legislators also cited the need to protect future sources of water supply and safeguard other natural functions. With the authorization of the Catskill system, legislators followed through on the plan to tap these rural areas for their abundant waters. The burgeoning population of metropolitan New York needed rural landscapes not only for the escape they provided from the urban grind, but also for the natural resources that they contained. The Catskills, largely recovered from a previous round of aggressive resource extraction, were ideally suited to meet both these needs.[99]

CITY + COUNTRY

CHAPTER TWO

Up Country

Viewed from any perspective, the Catskill water system was a massive under-taking. It cost approximately $180 million (more than two billion dollars in 2012), almost tripled the amount of water delivered to New York City on a daily basis, employed thousands, molded landscapes, and took more than two decades to complete. Most of those associated with the project did not take this wide-angle view. For some it was merely a way to make ends meet; others viewed it as a chance to advance their careers. For thousands of European immigrants (mostly Italians) it represented their introduction to America; for African American mule drivers it was only the latest in a long line of public works projects. For those dis-placed from their homes it meant an end to their communities as they knew them.

The scale of the project went well beyond the construction of two reservoirs, two tunnels, and a ninety-two-mile aqueduct. The men who began surveying the Catskills in 1905 did not know it, but their work launched six decades of water-works expansion by New York City. The construction of the Ashokan Reservoir and the Catskill Aqueduct—the two major components of the first phase of con-struction—established a template of thought and action that guided New York's water supply strategy for much of the twentieth century. A detailed examination of the construction's first stage illuminates the technical, political, and social di-mensions of subsequent expansion of the water network.

The challenge of building a new water system was as much bureaucratic as tech-nological; constructing dams and aqueducts of the magnitude and complexity en-visioned by the city's engineers required an exceptionally capable administrative infrastructure. The formation of a highly professional and motivated organization dedicated exclusively to building a new water network was critical to the project's success. The new agency, the Board of Water Supply (BWS), did more than draft blueprints and oversee construction. It quickly became the most powerful and per-suasive voice in favor of further expansion of the supply network, dictating the city's water policies for more than five decades. The board's cultivation of a distinct identity was a major factor in its success. Tracing the development of the organization is thus critical to understanding the evolution of New York City's water network.

The city practiced democratic resource imperialism. With the sanction of higher political authorities, it appropriated the resources of the countryside to sustain its own growth. New York's weapons of choice were shovels, surveyors, engineers, and lawyers. By welding modern statecraft with engineering prowess, New York City remade nature for its own ends. But dominating nature entailed dominating people. As anthropologist James Scott has observed, many of the twentieth century's most harrowing instances of human tyranny involved the appropriation of natural resources.[1]

The legal and political norms of American governance tempered the city's excesses. Catskill residents complained vehemently that New York disregarded their rights and property, treating them like petty obstacles to be overcome, not human beings worthy of respect. Nonetheless, the state government passed legislation guaranteeing watershed residents their day in court. These proceedings were far from perfect—many Catskill residents found it difficult to navigate modern legal and evidentiary requirements—yet the process provided those most directly affected by dam and reservoir construction with a modicum of social and economic justice. On balance, waterworks construction harmed the local economy by obliterating boardinghouse tourism and removing fertile agricultural land from production. Many local men did, however, obtain steady employment as laborers on the Catskill project.[2]

Construction of the Ashokan Reservoir, which began shortly after the board was established in 1905 and was completed in 1915, marked the beginning of the fraught relationship between Catskill residents and New York City. The city repeated the process of land seizure and waterworks construction detailed in this chapter several times from 1905 until 1967, when it completed work on the Cannonsville Reservoir. Unable to quench its imperial thirst, New York gradually extended its reach throughout the Catskills. Throughout this period, New York redeployed the political and technological strategies it relied on to build the Ashokan to maintain the upper hand in its relationship with mountain residents. Cast in the role of reluctant bride, Catskill residents sought to make the best of a marriage they played no part in arranging.

Building an Organization

In late June 1922, with one reservoir complete and another under way, engineers and the top brass of the BWS gathered in New York City for a testimonial dinner for J. Waldo Smith, to commemorate his retirement as chief engineer of the board.

Former mayor George McClellan, amazed at the scale of landscape change that Smith and his army of engineers and laborers had wrought in the Catskills over the past fifteen years, offered a bird's-eye view of these physical transformations, casting Smith in the role of magician:

> And such a fairy tale as it has been! Rivers turned backward from the sea, and made to flow through caverns in the hills, towns moved bodily from valleys to the heights, forests transmuted into lakes, and a great sea called into being where hills and valleys used to be. Overnight an almost desolate countryside made populous with thousands of the magician's men summoned from distant lands to work his will. And overnight again the surface of the earth is changed and mountains disappear and woodlands vanish and water is where dry land was, and once again man has triumphed over nature.[3]

McClellan's fairy tale version of history glossed over the sacrifices made by watershed residents whose lives and communities were irrevocably altered by New York's waterworks development; the tedious, sometimes deadly tasks required of the project's laborers; and the political conflicts that at times threatened the very viability of the massive construction project. Such messy realities had no place on this celebratory evening. McClellan's self-aggrandizing tendencies irritated many New Yorkers, but all present that evening agreed that the Catskill project represented "the concrete expression of the personality and of the genius of Waldo Smith."[4]

An unassuming man—the anti-McClellan—Smith appeared to lack the aggressive demeanor required to successfully navigate the city's turbulent politics. After graduating from the Massachusetts Institute of Technology in 1887, Smith left New England for New Jersey, where he spent more than a decade developing waterworks for the state's burgeoning cities. In 1903, he crossed the Hudson to take the position of chief engineer of New York City's notoriously inefficient Aqueduct Commission.[5] Despite his success at the commission, Smith lacked a national profile and did not play an important part in the campaign to secure state permission to tap Catskill watersheds. He figured to spend the next few years overseeing construction of the two additional reservoirs New York planned for the Croton region, a significantly less glamorous prospect than constructing an entirely new water system. When more prominent engineers such as John Freeman rejected the post, McClellan and the BWS tapped Smith as chief engineer.

Smith's success rested as much on his leadership abilities as his formidable technical prowess. In 1907, a New York newspaper described his relationship with the BWS commissioners, his nominal bosses: "Smith has been practically czar of the board, but a czar whose rule is unobtrusive and quiet."[6] Smith's understated style appealed to his subordinates, who took to expressing their admiration for the man they called "chief" in songs they wrote about their experiences working on the project. At the annual dinner for the chief engineer and other board events, employees belted out tunes like "Order Another One, Waldo," which emphasized his everyman qualities: "each worker knew that the boss was true/blue, a good fellow for ever and aye."[7]

Smith's evenhanded reaction to an aqueduct-testing accident offers insight into his management style. After a worker failed to open a gate, leading to a buildup of pressure in the Catskill Aqueduct that propelled 1,500-pound concrete slabs thirty feet into the air, he took the high road, philosophically remarking, "Well, they had to learn what water will do."[8] A demanding but shrewd overseer, Smith saw little point in expressing his displeasure more forcefully.

Building a competent and energetic organization that would deliver the project on time and on budget ranked as Smith's most enduring legacy. The city hired contractors to build the dams, reservoirs, and aqueducts, but board employees designed and supervised all aspects of the project. By 1906, Smith had established six departments to oversee different aspects of the project, each headed by an engineer who reported directly to him. Smith gave his engineers tremendous latitude to select their employees, intervening only when absolutely necessary. According to Robert Ridgway, an engineer who oversaw construction of the northern portion of the Catskill Aqueduct, Smith's low-key approach produced a "wonderful morale and esprit de corps in the force."[9]

The sheer magnitude of the undertaking made forming a cohesive organization difficult. At the height of construction, the BWS employed over two thousand people dispersed over several counties; the six engineering departments were in turn subdivided into sections focused on discrete tasks, such as building a pressure tunnel or clearing a future dam site of vegetation. Smith and his top engineers sought to cultivate pride in the organization as a whole as well as instill a powerful group identity among its individual units. The *Catskill Water System News*, a biweekly publication distributed by BWS headquarters in New York that relied on reports from employees in the field, exemplified this dual approach.

The *News* played a critical role in bolstering employee morale and creating a shared sense of purpose among its geographically dispersed workforce. The paper published notices about upcoming board events, updated workers about the progress of construction, and gave employees of one division an opportunity to communicate with those laboring forty or fifty miles away. After easily defeating the baseball team from Section 5 in July of 1911, the team from Section 2 used the paper to convey a challenge to "any team in the Southern Aqueduct department."[10] Dispatches from employees stationed in more remote areas read like letters from summer camp. In November 1912 an engineer at the Ashokan Reservoir informed distant co-workers that "the monotony of life at Brown's Station was interrupted" by a Halloween party complete with a cabaret performance.[11]

BWS management relied on the *News* and annual events such as the interdepartmental bowling tournament and "Field Day," which featured swimming competitions and baseball games—topped off with a clambake—to unite and motivate thousands of scattered employees.[12] For those laboring in remote areas, such events provided a diversion from the daily grind and a pleasant venue for catching up with co-workers stationed far down the aqueduct line. Thoroughly urban in both origin and mind-set, most engineers found rural life dreary, and longed to hear "the trolleys humming/buzzing night and day."[13] Like many large employers in the Progressive Era, the board offered its workers various programs intended to address their social needs and boost morale. The social programs of welfare capitalism—a staple of industrial firms seeking to pacify laborers considering unionization—helped knit together the board's workforce.[14]

Constructing massive dams and the world's most technologically advanced aqueduct to carry water almost a hundred miles from the mountains to New York City also galvanized board employees. High-level engineers such as Robert Ridgway appreciated the opportunity to contribute to such an ambitious and innovative project. Ridgway, whose long career included a stint at the Aqueduct Commission and decades overseeing subway construction in New York City, remembered his years at the BWS as "very fruitful and happy ones. The romantic nature of the work and the magnificence of its operation appealed to me strongly."[15]

Building a New Water System

Even to a seasoned engineer like Ridgway, the scale and complexity of the proposed water network were awe-inspiring. Capturing water and delivering it to New Yorkers entailed a massive rearrangement of human and physical geography. The board and its contractors needed to assemble a substantial workforce to build the new water system's dams, aqueducts, and tunnels. Laborers relocated to the Catskills and along the line of the aqueduct to take advantage of new employment opportunities. As their work progressed, residents whose homes would be flooded or whose business prospects faded began to leave the construction area in search of new lives and livelihoods.

The project was a complex combination of adding and erasing. The most significant addition was the massive Ashokan Reservoir, which was formed by damming Esopus Creek at Olive Bridge. Constructed from 1907 to 1915, the Ashokan was the principal water source for the project. It flooded several thousand acres,

38~ different dates

erased eight towns, and forced two thousand people to relocate. Ashokan's waters flowed over seventy miles via the Catskill Aqueduct to a revamped Kensico Reservoir near the Westchester County community of Valhalla. The original Kenisco, constructed in the 1880s, dammed the Bronx and Byram Rivers, providing the city with a small supplemental water source. Its much larger replacement submerged the original dam, flooding three communities and displacing five hundred residents. The new reservoir blended small amounts of water from the Bronx River with substantial flows from the Ashokan. After leaving Kenisco, mountain water flowed through the final section of the Catskill Aqueduct to Hillview Reservoir in Yonkers, on the Bronx border. Far smaller than the Ashokan and Kensico, Hillview regulated the flow of water into the delivery system, releasing flows during the day when demand was heavy and replenishing supplies at night. From Yonkers, water flowed through City Water Tunnel No. 1, an enormous cement-lined pressure tunnel that ran several hundred feet beneath city streets and connected to water mains throughout New York.[16]

Although the most innovative engineering accomplishments occurred below-ground, the visual grandeur of the new water system lay in its impressively designed dams and reservoirs. Ashokan Reservoir, with a capacity six times that of the recently constructed New Croton Reservoir, overshadowed virtually all other aspects of the project. Early accounts and representations of Ashokan focused on its remolding of the Esopus Valley and the scale and nature of the construction process. By 1915, when water began to back up behind the dam, Ashokan had come to represent not the triumph of man over nature, but the seamless blending of the two. The clearest evidence of this shift in perspective is the changing subject matter and style of postcard depictions of the Ashokan. Early postcards generally featured drab photographs of dams and dikes. As the project neared completion, postcard manufacturers abandoned their documentary-style infrastructure photos in favor of more artistic (and colorful) renderings of the reservoir. A 1916 postcard featured a lone boat bobbing gently on the reservoir's placid waters on a moonlit night. By the 1930s, even those images nominally centered on manmade features of the reservoir emphasized its naturalistic qualities. A 1935 postcard of the bridge that divides the reservoir's two basins depicted a man who has stepped out of his car to take in the view, the bridge less a triumph of engineering than a vehicle for the higher enjoyment of nature. The foreground of the postcard, bedecked with greenery, further reinforces the idea that the appeal of the reservoir lies less in its dams and dikes than in its intimate intermingling with its natural surroundings. A caption to a local photographer's series of reservoir images from the 1980s articulates the logic animating both his own work and that of the postcard illustrators: "The Catskill Mountains and the Ashokan Reservoir together create one of the most scenic and panoramic combinations depicting the work of man and the forces of nature."[17]

Fig. 4. Postcard, "Ashokan Reservoir by Night." This romantic postcard of the Ashokan Reservoir was produced while the reservoir was still filling. Recreational boating has never been permitted on the reservoir.

Fig. 5. Postcard, "Arch Bridge at the Dividing Weir, Ashokan Dam." This mid-1930s postcard depicts the easy intermingling of the man-made and the natural.

The construction process also involved a curious blending of natural and man-made forces. Equipped with derricks, railroads, and cableways, chief contractor Winston & Co. deployed the most advanced technology available. Nonetheless, contractors relied primarily on mules to haul cement bags and other construction materials, and on occasion used them to remove tree stumps from the reservoir floor.[18] African American men managed the mules, and often traveled from one public works project to another, using old-fashioned animal power to build technologically sophisticated dams and bridges. Southern blacks drew on their agricultural heritage to build critical urban infrastructure for northern cities. In an era before powerful trucks, mules were crucial to the success of the project, as evidenced by the careful tracking of the animal population at each work site, which at times approached almost a third of the human workforce.[19] The mules were a source of amusement to children, who loved to watch teamsters race their beasts down the roads at quitting time. The local farmers who fed and stabled the animals welcomed the additional income they provided.[20] Local materials also proved vital to the project's success, particularly the abundant seams of bluestone that underlay large portions of the Catskills. Winston & Co. combined bluestone taken from a quarry site three miles from Olive Bridge with concrete blocks to build the cyclopean masonry of the reservoir's dam.[21]

A wide range of legal and practical requirements forced BWS and other city officials to interact closely with those who worked on and lived near construction projects. The broad geographic scope of the project required contractors to establish dozens of work camps to house and feed the largely immigrant workforce. Acting under strict orders from the board, contractors created camps that clearly reflected the values and biases of the time. Although in many respects the camps represented vast improvements over the squalid conditions that often prevailed at rural work sites (and over the insalubrious tenements where most recent immigrant arrivals to New York City lived), they also firmly reinforced prevailing racial hierarchies. Italian and black workers generally occupied separate quarters at the sprawling camp below the dam of the Ashokan Reservoir, while "white Americans lived separately."[22] The family status of camp residents occasioned further segregation; married men generally lived in four-room houses equipped with kitchens and porches, whereas single men resided in dormitories or barracks and cooked for themselves. Segregation extended to the workplace as well, where most African Americans worked as mule drivers, while some Italians received higher wages as expert stonecutters.[23]

A frequent summer visitor to the Catskills observed in 1910 that the establishment of the main camp for Ashokan Reservoir workers had transformed Brown's Station from "a quiet village to a busy town of 2500 inhabitants."[24] Winston & Co. created a company town atmosphere at its main work camp, which included a general store, bakery, post office, churches, a restaurant, and schools. The company's efforts to sequester its workers in the camps reflected both a desire to profit from their purchases and to minimize conflicts with surrounding villages. In many instances the line between the work camps and local villages proved quite porous. Laborers patronized village stores on their rare days off; Winston & Co. sold some of the food it purchased for its workers to local residents; and scores of local men, attracted by the steady wages, helped build the new reservoir.[25]

Building the dams, dikes, aqueducts, and reservoirs needed to store and transport water from the Esopus required an enormous workforce. The population of the town of Olive, where most of the men building the Ashokan Reservoir lived, almost doubled with the influx of workers.[26] Corralling their wastes before they contaminated local streams ranked as one of the project's greatest challenges. The *health* BWS fenced most work camps, installed incinerators to burn garbage and fecal matter, and thoroughly treated liquid wastes before discharging them into watercourses. Contractors agreed to supply workers and their families with pure drinking water, and to establish isolation wards to quarantine workers suspected of carrying communicable diseases.[27] The relatively low death rates in the camps—on average, three and a half camp residents per thousand died in a year, exclusive of accidents, a rate well below comparable public works projects—attested to the effectiveness of these precautions.[28]

Typical of its overall approach to the project, the board relied on a combination of technology and manpower to ensure the purity of the water it collected in the Catskills. It imposed a highly regimented program of sanitary control over both work camps and the sprawling watershed. To limit potential disease outbreaks, the BWS required camp physicians to file weekly health reports detailing the number of patients treated and the nature of their illnesses. It supplemented this reporting with aggressive patrolling of work camps. To the tune of "Yankee Doodle," project engineers sang of the sand filters they designed for the camps and of the complementary role their police brethren played in ensuring water quality:

> The cop's around the camps at night
> To seek some information,
> Of the men who violate,
> The laws of sanitation.[29]

Workers were encouraged to spend their free time at the camps, where BWS police and contractor staff could monitor their actions and wastes.

Controlling sanitary conditions throughout the Esopus watershed proved much more challenging than limiting pollution in the relatively circumscribed world of

pollution control v.
filtration

the work camps. The board worried that Catskill residents, despite their low numbers, posed a serious threat to water purity: "Numerous outhouses are located so close to the streams as to require their immediate removal, while factories, as well, discharge spent chemicals and other waste matters directly into the channel."[30] The city sought authority from the state to remedy such conditions. In doing so, it implicitly acknowledged that the Catskills was not virgin territory; the region would have to be scrubbed clean to ensure the purity of New York City's water. After two failed attempts, in 1915 the city secured passage of state legislation permitting it to establish and enforce health standards in the Catskill watershed. The resulting regulations outlawed disposal of garbage, manure, and other wastes into watercourses; banned privies, stables, and similar structures from a buffer zone around streams and reservoirs; and prohibited laundry washing and swimming in the reservoirs. In the short term, the regulations led to new outhouses and privies. By the 1920s, New York City had begun to construct sewage treatment plants in Catskill villages to protect its substantial investment in the purity of mountain water.[31]

The aggressive campaign against pollution in upstate watersheds stemmed in part from the city's reluctance to filter its water supply. The original plans for the Catskill project included a $17.5 million filtration plant, leading New York to acquire land in the Westchester County town of Mount Pleasant for a facility.[32] In 1909, the board appeared to commit to filtering the Catskill supply: "The public demand for water of a high standard, both in appearance and in purity, was recognized at the inception of the work by making provision for a full-capacity filtration plant in the original plans."[33] This rhetoric clashed with the city's refusal to pay for filtration. In 1908, municipal officials canceled plans to construct a filtration plant for supplies drawn from the Croton region, whose water was far dirtier than the new Catskill supply. By 1915, with the delivery of unfiltered Catskill water only months away, consulting engineer John Freeman "could see no reason why this water cannot be prudently delivered for some years without filtration, if the present financial needs of the City require deferring such expenditures."[34] Once the state granted New York the authority to impose sanitary regulations on Catskill watersheds, the incentive to filter the new supply quickly disappeared. Replacing privies, constructing small sewage treatment plants for Catskill villages, and paying the handful of board policemen charged with monitoring local residents' compliance with sanitary regulations was much less expensive than building a filtration plant. The question that would come to dominate New York City's relationship with Catskill residents in the 1990s—should it build a costly filtration plant to improve water quality or instead enact strict regulations to reduce pollution on the watershed—first reared its head, albeit less starkly, seventy-five years earlier. Rather than extract more revenue from its citizens to build a filtration plant, the city sought to indefinitely defer filtration by imposing its police powers on watershed residents.

Although it took almost a decade to complete, the Ashokan nonetheless proved relatively straightforward to build compared to the difficulties encountered in designing and constructing the aqueduct that would carry the reservoir's water to New York. On its route between Ashokan and the Kensico Reservoir, the aqueduct "skirts along many a steep hillside, pierces mountains, descends beneath rivers and wide, deep valleys."[35] Where land was flat or sloped gently, engineers generally called for contractors to dig a shallow trench in which to submerge the aqueduct, a technique known as cut-and-cover. The most difficult sections to construct were the pressure tunnels, those portions of the aqueduct that dipped hundreds of feet below the earth's surface. Built of steel and steel-reinforced concrete and encased in hundreds of feet of solid bedrock strong enough to absorb the force of the water, pressure tunnels were among the most time-consuming and difficult aspects of the entire project because of the expensive materials employed and the extensive borings required to penetrate hundreds of feet below the earth's surface.

Devising a means to carry Esopus water beneath the Hudson River was the most challenging aspect of constructing the aqueduct. Considered too polluted and brackish to serve as a water source, the Hudson was the most formidable barrier to conveying the pristine waters of the Catskill Mountains to millions of thirsty New Yorkers. Engineers identified two places where the river narrowed as suitable crossing locations, ultimately settling on the area around Storm King, a few miles south of West Point. In contrast to the Catskills, where the BWS insisted that the Ashokan Reservoir would greatly enhance the region's beauty, erecting a bridge to carry the aqueduct over the river threatened to mar the spectacular scenery at one of the Hudson's most picturesque locations. In Robert Ridgway's words, "The spanning of the channel with a bridge would be regarded by many almost as a sacrilege."[36] He may have been right, but it was economics (digging a tunnel was much cheaper than building a bridge), not aesthetics, that ultimately convinced the BWS to go under rather than over the Hudson.

Building a mile-long pressure tunnel to convey Catskill water under the river presented a unique set of challenges. Boring into the earth below the riverbed to locate bedrock thick enough to withstand the enormous pressure exerted by the water inside the aqueduct proved practically impossible. Workers used diamond drill bits to plumb the Hudson's depths, but these bits made agonizingly slow progress, sometimes only inches a day. After tides and traffic dislodged several borings before they could penetrate more than a few hundred feet, engineers decided to take borings from the banks of the river. After finding solid rock at 955 feet below the earth's surface, engineers called for building the tunnel 1,100 feet below the Hudson to ensure a cushion of at least 150 feet of bedrock to stabilize the aqueduct.[37] In

song

what would become a common practice among board employees, the engineers celebrated their achievement with song: "We fought bedrock for ages and finally got a clew / And down at the 'leven hundred we're putting the siphon through!"[38]

After coursing under the Hudson, Catskill water flowed to the Kensico Reservoir, the architectural showpiece of the new water system. The board worried that the entire Catskill supply could be disrupted if the aqueduct failed at some point during the long journey from Ashokan to New York. Establishing a large emergency reservoir in Westchester County would provide New Yorkers with several weeks of water if disaster struck, and offer the added benefit of improved water quality: the longer water sat in reservoirs, the more time currents and the sun had to filter out impurities.[39] A collection of mostly local workers labored for six years to build a new reservoir at Kensico. Engineers conceived of Kensico—located a mere fifteen miles from the New York City line—not as a remote rural reservoir, but as a suburban destination for the increasing number of urban dwellers who owned automobiles. The Bronx River Parkway would stretch over fifteen miles from the Bronx Botanical Gardens to a reservoir park located just downstream from Kensico's main dam on the Bronx River. The board noted approvingly that the connection from the parkway ensured that "the most monumental structure of the Catskill system will have a suitable approach and setting."[40] City and suburban drivers packed the parkway, seeking out the reservoir's expansive grounds on weekends and summer evenings. Historian Barbara Troetel recalled her father "driving the family up the Bronx River Parkway to the Kensico Dam on a hot summer evening. . . . We would sit on the grass in the cool darkness near the Kensico fountains until late in the evening." Although the reservoir it created was only a quarter the size of Ashokan, Kensico's three-hundred-foot dam loomed fifty feet higher than the one at Olive Bridge.[41]

Kensico's grandeur was enhanced by the decorative elements that adorned the dam's downstream face and top. Faced with local granite from a nearby quarry, the dam, with its rough-hewn appearance, appealed to contemporary observers. Fifteen-foot-wide bands of rusticated stone projected from the dam's face at fixed intervals, dividing it into a sequence of curving stone sections and lending architectural variety to the massive wall of stone. A parapet provided impressive views of the rectangular pool and fountains just beyond the dam, and of the surrounding countryside. An updated version of the Distributing Reservoir at Murray Hill, Kensico blended the ornamentation of the City Beautiful movement with the sleekness of modernist architecture.

Consulting engineer John Freeman originally viewed these details as extravagances that unnecessarily increased the project's cost, but by the time the dam was completed in 1919, stands of trees had grown up amid the fountains, and reflecting pools and carefully laid out walkways had turned Kensico into an architectural

Fig. 6. Kensico Reservoir, c. 1918. East pylon, fountain, and cascade basin show some of the "extensive decorations" incorporated into Kensico's design. (Science, Industry and Business Library, General Collection Division, New York Public Library)

showpiece. He stressed the informed quality of his praise: "I have seen most of the notable high dams of the world . . . and it is my profound belief that this is the most beautiful dam in the world."[42] Other observers agreed with Freeman. Desmond Fitzgerald, an engineer who played a critical role in expanding Boston's water supply and had toured numerous dam sites, told his fellow waterworks designers, "the architect . . . had an unlimited order. I never have seen anywhere in the world more elaborate—well I won't call them "elaborate" because they are sufficiently simple, but extensive—decorations."[43]

Instead of constructing a monumental structure intended to capture the public's imagination, tunnel engineers were charged with designing an invisible water delivery system. Geology and economy led the BWS to build the world's longest pressure tunnel expressly designed to convey water; the eighteen miles of City Tunnel No. 1 stretched under three boroughs, crossing two rivers and New York Harbor. The exceptionally thick layer of bedrock that underlay much of the city absorbed the pressure of rushing water.[44] Because the tunnel was submerged hundreds of feet below street level, it did not interfere with subway tunnels or other utilities. Unlike steel pipelines, which had to be replaced periodically, a concrete pressure tunnel would last for decades with virtually no maintenance.[45]

More than any other component of the Catskill water system, City Water Tunnel No. 1 brought into sharp relief the urban sensibility that characterized the entire project. The construction of a water distribution system that minimally altered daily life in New York and that resulted in only one aboveground structure within the city (Staten Island's Silver Lake Reservoir), contrasted sharply with the substantial social dislocation that occurred in the Kensico and Ashokan regions.[46] BWS engineers believed that their reservoirs would beautify and elevate the roughness of the countryside. The city, where skyscrapers had recently begun to redefine the horizon, required no such architectural makeup. Board employees did regret that creation of the Ashokan Reservoir destroyed Bishop Falls, whose dramatic cascade and picturesque mill on Esopus Creek had made it a favorite relaxation spot for overworked engineers, but their lament for the falls was a notable exception.[47] Intent on improving rural New York by sculpting scenic landscapes, the board buried most of its urban infrastructure. Even as the visible imprint of water infrastructure faded in the city, the BWS looked to construct more dams, reservoirs, and tunnels in the Catskills.

The View from Gotham

Board and municipal leaders maintained a remarkably consistent attitude regarding the fundamental aspects of the project. They never wavered from the belief that their primary duty was to continually expand the supply network to provide New Yorkers with ever-increasing amounts of high-quality water from the Catskill Mountains. Restraining the seemingly inexorable growth in water consumption played no part in this strategy. Equal parts utilitarian and utopian, this vision encompassed the construction of magnificent structures and the provision of an essential public service to millions of New Yorkers. To realize this vision, the board needed to modify nature and to regulate the use of natural resources in the Catskills. The notion of a managed landscape reflected both the prevailing conservation ethos of the era articulated most forcefully by U.S. Forest Service chief Gifford Pinchot, and the need to control pollution in Catskill watersheds.[48] This program of controlled extraction and the health regulations and policing of watersheds that accompanied it irrevocably altered the lives, livelihoods, and landscapes of thousands of watershed residents.

For political and psychological reasons, the city portrayed the Catskill watersheds as wilderness areas with low permanent populations. Sacrificing what McClellan had disparaged as "this almost desolate countryside" to provide water for millions of New Yorkers struck most urban observers as a plan with no obvious drawbacks. Just as English colonists romanticized America as a wilderness almost wholly untouched by human influence, and settlers overlooked the aridity of the West when they proclaimed the region America's garden, New Yorkers reduced the Catskills to a watershed in waiting.[49] In a statement approving the city's application to develop the Esopus watershed in May 1906, the State Water Supply Commission echoed this sentiment: "The mountain region is not liable ever to become a manufacturing district or the center of a large resident population and is therefore peculiarly adapted for a permanent watershed."[50]

This narrow reading of the region obscured the intimate connections binding city and country. New Yorkers had long relied on the region's natural resources. By paving its sidewalks with bluestone, New York City helped transform the physical and social landscape of Ulster County decades before it began to tap its waters. Dark bluestone from the Catskills proved the ideal material for city sidewalks because it resisted polishing and thus never became slippery. Its low cost and durability enhanced its appeal.[51] The rise of the bluestone industry in the late nineteenth century led to significant churning of the population in northern Ulster County, as workers abandoned depleted quarries in search of more profitable seams. As A. W. Hoffman observed in his survey of the industry, "Places that once flourished are now in decay and other places have grown up."[52]

The Catskills also offered New Yorkers another commodity in short supply: fresh air. By the 1880s, several Ulster towns, including Hurley and Olive—the two communities most directly affected by construction of the Ashokan Reservoir— had become tourist retreats for middle-class urban residents. Families often spent several weeks, and in some cases up to two months, enjoying the natural beauty of the Catskills. In 1904, the influx of summer visitors sustained four hotels in the small village of West Hurley, and general store owner P. M. Barton ran a de facto hotel in his expansive home, often hosting up to thirty visitors at a time.[53] Most of the farm families who patronized the store hosted boarders in the summer months, relying on the income they earned from tending to the needs of these urban dwellers. The extension of the Ulster and Delaware Railroad in the 1870s made the region accessible to New Yorkers, who took steamers up the Hudson to Kingston, where they connected to U&D trains that carried them into the Esopus Valley. In 1870, only those hardy souls with the time and gumption required to navigate a sparse and poorly maintained road network visited the Catskills; by 1883, an estimated two hundred thousand visited the mountains each year, turning it into a major tourist destination.[54] Almost overnight, communities near a railroad depot found their economies transformed by the urbanites who packed private homes in search of country air and hospitality.

City officials consistently downplayed the impact of waterworks construction on the Catskill landscape and economy, focusing instead on the project's place in the larger sweep of history. Transfixed by that place in history, boosters often emphasized the past over the present. The Catskill system naturally invited comparisons to the construction under way on the Panama Canal, but those associated with the project preferred to highlight its connections to its aqueous ancestor, Rome, and other ancient civilizations. Sympathetic observers cast New York as Rome's contemporary disciple, a modern power intent on building the ultimate democratic metropolis. At a 1907 ceremony marking the official beginning of construction, BWS president J. Edward Simmons compared the Catskill water project to the public works of Babylon and Rome. New York's project dwarfed Rome's in size, he observed, but would be built by free men, not slaves. "A colossal monument to the achievements of modern science and intellect," the Catskill project represented the apogee of democratic modernism: conceived and designed by engineers, it would benefit the masses drawn in ever-increasing numbers to New York City.[55]

Project employees and boosters also emphasized more distant historical connections. Geological surveys commissioned by New York revealed that the site of the Ashokan Reservoir had been a lake before glaciers deposited their till, creating the valley. Many, including travel writer T. S. Longstreth, presented the construction of the Ashokan Reservoir not as a victory of man over nature (the conventional martial analogy adopted by McClellan), but as a reunion with it. He rhapsodized, "The lake was so beautiful, fitted so well into border—land of mountain and plain, that it did not look raw and new. To tell the geologic truth, it had been on the original plan of the globe. The surveyors found evidences of a pre-glacial lake. All they did was to put it back."[56] Highlighting the reservoir's geological history was one means of naturalizing reservoir construction, a rhetorical strategy that neatly elided what had transpired in the meantime: permanent settlement by human beings who treasured the valley for its natural resources, beauty, and emotional connections.

When the city did design to acknowledge the humans who lived in the valley, it focused on Indians, not the current inhabitants. In an allegorical pageant written for the 1917 celebration marking the completion of the Catskill Aqueduct, New York's water commissioners asked the Indian chief Ashokan for permission to divert water to the city.[57] Ashokan graciously consented, and the mayor proceeded to sprinkle water on each of the five boroughs. The present-day residents of the Catskills received no role in the pageant.[58]

Economy and Culture of the Catskills

At the end of the nineteenth century, the economy of central Catskill communities such as Olive and Hurley rested on three basic pillars: natural resource extraction, tourism, and agriculture. Many men changed jobs with the season, logging in winter, working at the quarries when they opened in the spring, and cultivating their fields during the growing season. Women were busiest in the summer, when most opened their households for twelve weeks to boarders who, for seven to nine dollars a week, made the Catskills their second home. Although the accommodations were often quite spartan—families typically slept four to a room and dined shoulder to shoulder with twenty other guests—visitors enjoyed the fresh farm produce and the orchards, streams, fields, and mountains that surrounded their lodgings. Even in fertile river valleys, many residents found it difficult to sustain their families by farming alone. Raymond Carruthers, who spent his childhood in West Shokan, one of Olive's seven villages, recalled that the income earned from boarders allowed families to remain on the farm.[59] New York City comptroller Herman Metz took a slightly dimmer view of Catskill farmers, concluding, "The New York vacationist has been cultivated more assiduously, perhaps, than the fields."[60]

Metz accurately assessed the boom in the boarding business—by 1908 over one thousand hotels and boardinghouses in Ulster and Delaware Counties welcomed summer guests—but he failed to appreciate the symbiotic relationship between agriculture and boarding.[61] Instead of abandoning agriculture for boarding, farmers reoriented their sowing, reaping, and animal husbandry to maximize their profits from boarders' weekly payments. The more food they could raise from their own lands, the less they would have to purchase from local butchers and general stores. The eggs and butter that Martha Young fed the twenty-odd guests in her West

Hurley home came from chickens and cows she kept on her farm.[62] Clara Gricks, despite growing her own vegetables and raising chickens, estimated that she spent almost half her income purchasing meat, grain, and other food for her boarders.[63]

From the perspective of urban business owners, the economics of the boarding industry made little sense. As the city's attorneys frequently observed, after factoring in wages for those who fed and cared for visitors (who were usually family members and therefore did not generally receive wages) boardinghouse operators turned little or no profit. In contrast, Harrison Slosson, a Westchester County attorney who represented hundreds of Catskill residents seeking compensation from New York City, vigorously challenged the notion that Esopus Valley residents led a hand-to-mouth existence, arguing that "almost every farmer in that whole section kept boarders during the summer season, and from those boarders they were able to make a good living."[64] In fact, boarding generally represented more of a rural survival strategy than a means of capital accumulation. As city attorney William Speer observed, by operating a boarding business, valley residents ensured a ready market for their farm produce. At the end of the boarding season, most families had enough money remaining to buy clothes, pay their taxes, and cover other essential needs. Boarding may not have provided a lucrative income, but it did allow Catskill residents to remain on their land, a feat many northeastern farmers found increasingly untenable.[65]

The affection that valley residents had for their homesteads embodied more than simple love of fields and houses. They also valued the small-town charm and conveniences of the region's villages. These villages and the roads and railroads that connected them gave rural life a decidedly communal and, in some cases, sophisticated flavor. Ernest Hamlin Abbott, a New York City journalist who visited the valley in the early stages of reservoir construction, came away impressed with the tidy hamlets. He was especially enamored of West Shokan, with its "homelike village street, bordered by well bred houses which seem to have nurtured one generation after another of sound and well born Americans."[66] West Shokan native Elwyn Davis referred to the village as the "Eden of the Catskills" for its tree-lined streets, lovely homes, and prosperous businesses.[67] Attracted by its railroad depot and well-stocked general store, hundreds of boarders spent summers in West Shokan. In the afternoon, visitors and locals met in the general store to collect their mail and make phone calls; at night they packed the village's three-hundred-seat Pythian Hall, where they enjoyed plays and musical performances.[68]

Of course, boarders came to the Catskills to escape urban life, not just to reproduce it on a smaller and more human scale. The Ulster and Delaware Railroad, which maintained several stations in the Catskills, emphasized the region's regenerative qualities in its promotional literature: "A breath of Nature, uncontaminated by the dregs of city civilization, is the unfailing panacea. The flabby muscles and pale cheeks, the feeble respiration and the exhausted brain, all these beckon us away to the green hills and valleys."[69] For most visitors, the surrounding mountains were a pleasant backdrop, not a place to thoroughly explore. They spent most of their time in the valley, relaxing under shade trees or by the many creeks and streams. One local farmer recalled finding a group of boarders plucking apples from his trees. He angrily confronted them, asking how they would react "if I went to one of your stores and started taking apples without paying for them?"[70] The incident no doubt reflected urban dwellers' ignorance of the economic fundamentals of rural life, but it also suggests boarders' desire to immerse themselves in the idyllic surroundings of these quiet villages.

The City Cometh

Waterworks construction marked yet another economic transformation for the Catskills. Dams and reservoirs reshaped the landscape, and in the process undermined boarding and agriculture, the economic mainstays of the region. Although much of the damage inflicted by New York City's thirst for water was communal—reservoir construction flooded entire communities, and many residents moved outside the region—the process of compensating residents for their losses happened one home and business at a time. State law entitled New York City to take possession of privately owned properties to build a water system, but it established a clear legal process to ensure that property owners received fair compensation for their holdings. As a dam began to rise on the Esopus, people gathered in courtrooms throughout southeastern New York State to debate the value of homes and businesses that the city was busily tearing down.

The construction of the Ashokan Reservoir introduced an entirely new segment of New York City's population—its lawyers—to the Catskills. Their primary task was reducing the cost of acquiring the thousands of acres required for the project. To ensure that residents had enough to live on while their cases awaited final disposition, state legislation required New York to pay property owners half the assessed value when it acquired title to land. The city, under the auspices of the New York Supreme Court, was then required to convene a real estate commission consisting of three members (one from the county where the property was located, one from New York City, and a third commissioner from elsewhere in the state), which, after hearing testimony from New York City and the landowner, recommended an award amount. After approval of the awards by the court, the city made final payment to property owners. A similar process governed awards for indirect damages, payments made to compensate business owners and wage earners for reduction in income due to waterworks construction.[71]

Lawyers representing New York (some were municipal employees, others were Catskill attorneys hired for their familiarity with local customs and real estate) had a distinct perspective on the Catskill landscape. Focused on land fertility and proximity to railroad stations and town centers—factors that directly influenced the value of real estate—these attorneys faced off against local experts hired by property owners (known as claimants). City attorneys deployed their unique talents to minimize the financial impact of land taking on New York's coffers. In their insistence on standardized data and quantitative evidence, the lawyers spoke what historian Daniel Rodgers has called the "language of social efficiency," a hallmark of Progressive Era ideology. Their commitment to a fixed set of standards by which to evaluate thousands of property claims reflected the desire for an efficient means of resolving cases. In attempting to make the welter of individual cases "legible" to the presiding commissions, the attorneys undertook the legal equivalent of standardized manufacturing of consumer products.[72]

The motivations of Catskill residents whose homes and businesses were seized or damaged because of waterworks construction cannot, of course, be rendered as neatly as those of outside actors charged with performing a narrowly defined task. Some locals undoubtedly relished the opportunity afforded by land awards to pull up stakes and seek a better life elsewhere. However, these adventurous souls appear to have been the exception, not the rule. Most of those displaced moved either to the new villages residents established to replace their old communities (these were located on the hills overlooking the reservoir) or to the nearby county seat of Kingston.[73] Determined to stay put, residents of the Ashokan region recognized the need to leaven their resentment of the city's intrusions with a healthy dose of pragmatism. Battling New York before a damage commission to extract maximum compensation for a doomed business in no way precluded a resident, or that person's relatives, from laboring on the waterworks project.

By the summer of 1910, with workers making steady progress on the Olive Bridge dam and several dikes, the physical, economic, and social tenor of life in Hurley and Olive had changed forever. Active construction scared off longtime summer visitors. Eugene O'Connell, a young New York City lawyer whose family typically spent several weeks in West Hurley each summer, explained why his father took his mother and sister back to the city after spending only three days in the Catskills the previous year: "Well, the Italians and colored people walking up and down past the road, different days and Sundays; there would be crowds of them going up and down through the woods, and he didn't want to leave my mother and sister around there, without some—he didn't have any time to stay up there himself just then."[74] Racist stereotypes, mingled with genuine concern about the safety of his family, led O'Connell's father and many other longtime boarders to seek a new summer retreat.

The onset of active construction erected economic obstacles that proved impossible for many longtime residents to overcome. Some of the two thousand residents whose property was condemned to make way for the reservoir solicited boarders for their new homes, often only a few miles from their old residences. The experience of Clara Gricks was typical. The convenient location of her former residence attracted many visitors; it sat at the junction of five roads, and boarders could walk to the West Hurley railroad station in a matter of minutes. Her new home lay in a remote area outside Kingston, far from any village center or railroad connection. Her son Rudolph, who lived and worked in New York City, tried to persuade his customers and friends to board at the new place, "but they objected to that quietude" and "wanted more access to life."[75] Uriah Wood, a butcher who spent his days peddling meat to customers in the reservoir area and beyond, also struggled to adapt to the region's new economic geography. With the commencement of active work on the reservoir in 1907, many of Wood's customers moved away. Those who remained lived on both sides of the rising dam, greatly increasing his travel time. Moreover, many of his former customers elected to purchase their meat from Winston & Co., which sold food to local residents at prices Wood could not hope to match. In the face of such daunting obstacles, the sixty-six-year-old chose to retire, ultimately receiving $1,300 for his butcher business in 1911.[76]

The economic consequences of reservoir construction were complex, sometimes cutting both ways within a single family. Winston's decision to sell meat to local residents ended Uriah Wood's butchering business, but the company also indirectly supported Wood by employing his son on the reservoir project. For Max Ferro, whose one-hundred-acre West Hurley parcel contained valuable pulpwood and bluestone, the temporary employment reservoir construction provided to his sons did not offset his diminished ability to earn a living from his land. Ferro's struggles illustrate the cascading effects that reservoir construction had on the valley's economy. Although the city did not take title to Ferro's property, it condemned the pulp mill where he sold his wood, depriving him of his only purchaser. Like most landowners in the area, Ferro relied on others to quarry his bluestone for him.[77] Eager to earn cash, if only for a few months and for modest wages, many local men had grown accustomed to spending part of their year extracting bluestone from the earth. While no less backbreaking, reservoir work proved more remunerative, leading Ferro to lament, "We could not get any men to work in the quarries."[78] The wood and stone that had once brought him decent returns now sat untouched. In 1920, he filed a claim for $1,800 against the city for indirect damages, hoping to recover some of his losses.

If the arrival of the reservoir dimmed business prospects for many valley residents, it created new opportunities for other locals. Well after the Ashokan's substantial completion in 1915, local men—and they were all men—continued to serve as expert witnesses before business damage commissions. Working for both New York City and the attorneys representing Catskill residents, these men testified in hundreds of cases, earning thousands of dollars in fees. Drawing on their expertise in farming, fruit growing, and building, they visited the parcels in question and offered informed opinions about the value of properties seized and businesses damaged by reservoir construction.

The city attorneys who handled the reservoir cases soon developed an easy, sometimes sarcastic, rapport with these expert witnesses, often asking them to explain apparent inconsistencies in their methods of property valuation in different cases. In one case, William Every, who frequently testified for claimants against New York, argued that the value of a West Shokan property was increased by virtue of its location on the north side of the street, which, due to prevailing wind patterns, rarely received the dust swirls common to homes on the south side. Skeptical of Every's claim, the city's attorney asked him why he had not raised this dust issue when testifying in cases about the value of homes on the south side. Every slyly deadpanned, "You didn't ask me."[79]

While Every spent his days sparring with attorneys in Kingston, many of the valley's men worked clearing the reservoir floor or building its dams and bridges. Fragmentary data make quantifying the number of local employees impossible, but frequent mention of local men working for the BWS or contractors in testimony before damage commissions and descriptions of the workforce on particular contracts indicate that many local residents worked at Ashokan, Kensico, and along the aqueduct line. At Ashokan, Catskill residents generally found work in projects ancillary to dam construction, such as removing vegetation from the valley floor or building the highways that would encircle the reservoir. At Kensico, locals actually outnumbered outsiders, reducing the need for work camps.[80] Men who worked on the project for several years moved up in the ranks, earning salaries previously unheard of for manual labor, especially during World War I, when labor shortages led to a spike in wages. Hired by the BWS in 1919 to assist with the second phase of the Catskill water project—construction of the Schoharie Reservoir and the Shandaken Tunnel—Ira Garlinghouse and John Davis initially received $3 a day. By 1920, Garlinghouse earned $4.20 a day, and Davis had been promoted to axman at an annual salary of $1,440. In 1922, Garlinghouse supplemented his salary by leasing some of his land to the board.[81]

Evaluating the Impact

The rise of Garlinghouse and Davis notwithstanding, reservoir construction in the Esopus Valley generated a wide range of individual and communal outcomes.

Determined to counter claims that the building of the Ashokan had sundered community ties by dispersing the local population, the board trumpeted the results of a 1913 survey indicating that over 80 percent of residents remained in the area. But the city's claims of community continuity were vastly overstated. Fewer than 40 percent of residents relocated to the area immediately surrounding the reservoir; fully a quarter, deprived of the income potential of their farms and boardinghouses, moved to Kingston in search of work.[82] This rate of out-migration resulted in significantly smaller communities. By 1920, Olive had lost over half its residents; the town of Gilboa, where Schoharie Reservoir construction began in 1920, shrank by a third.[83]

More important, the board's figures reveal nothing of the emotional toll that displacement from their homes and businesses took on valley residents. Eighty-three-year-old West Hurley native Joseph McElvey swore he would never leave his home. As his eviction date neared, McElvey wandered from his home, pistol in hand. He was found three days later, dead from exposure, with his gun by his side.[84] Although more abrupt than that of many of the neighbors he had known for decades, McElvey's death was one of many hastened by the extensive condemnation of area homes and businesses. Raymond Carruthers recalled that elderly people who rarely traveled more than a few miles from their homes lacked the emotional, social, and financial resources to cope with the torrent of change unleashed by reservoir construction. They often "worried themselves sick and died before their final payment for their property was received."[85]

Other residents channeled their energies into rebuilding their communities. They relocated their villages near the reservoir, generally maintaining the names of their original towns. Many buildings got jumbled and divided in a game of musical real estate: John Saxe converted West Hurley's Temperance Hall into a general store; Henry Barber fashioned a barn from the walls of George Winchell's house; and Lyman Smith built a new home from the remnants of Dr. Dumond's old one.[86] Residents relocated five of the eight villages submerged by the reservoir, reconstituting community life around new and rebuilt churches and general stores.

The remaining residents managed to eke out a living in their new hamlets, but the boarding boom seemed like a distant memory. By 1917, waterworks employees had departed and a hush settled over the region, leading boardinghouse owner Gene Kerr to complain, "Oh, if somebody would only come along and start something! Ever since the waterworks was started, the valley's been dead—no summer people, nobody to sell butter and eggs to. And it's a beautiful place, too."[87] Severe limits on recreational use of the reservoir and the lack of tourist infrastructure in the new villages hampered efforts to revive the boarding industry that had sustained the local economy. Visitors preferred to pass their days in northern reaches of the Catskills where they could enjoy upscale lodging and participate in a variety of outdoor activities. A 1902 Ulster and Delaware Railroad tourism guide touted the serenity of Bishop Falls, trout fishing in the Esopus Creek, and the "pleasant hamlet" of West Shokan. A similar guide published thirty years later focused almost exclusively on the lavish hotels and natural attractions of the Greene County Catskills, with the reservoir region rating no more than a brief day trip.[88]

With the boarding business dead, some residents hoped to sustain themselves by expanding their farming operations. One challenge was that their post-reservoir farms tended to be smaller and less fertile than their original lands.[89] In April 1913, a local magazine took stock of what would soon be submerged: "Many square miles of territory will disappear beneath the imprisoned waters and many acres of fertile farm lands, some of which have been cultivated for more than two hundred years, will be engulfed never to emerge."[90] Farm populations throughout Ulster County decreased in these years, but nowhere was the decline more precipitous than in the reservoir communities of Olive and Hurley. From 1875 to 1905, shortly before waterworks construction began, the number of Olive residents identified as farmers decreased by approximately 15 percent. From 1905 to 1930, with thousands of acres of farmland flooded to create the Ashokan Reservoir, the town lost over 40 percent of its farmers.[91] Adding insult to injury, prices for agricultural products fell steeply after World War I, making it impossible for many Catskill farmers to earn a living on their shrunken parcels.

The changes generated by waterworks construction were more wrenching and permanent than those from tanning, timbering, and bluestone extraction. Abundant rainfall ensured that the grandchildren of tanners and furniture makers enjoyed many of the visual and ecological benefits of lush forests. Unlike cutover forests, lands submerged by the reservoir would never restore themselves. Reservoir construction also characterized the rich layers of association and seasonal cycles of activity that underpinned valley life. In addition to submerging valuable farmland, the reservoir claimed much of the Beaverkill Swamp, a favorite spot for Hurley residents who enjoyed the swamp's spring riot of flowers and hunting partridge in its mists.[92]

If fishing on the reservoir provided partial compensation for the loss of recreational opportunities, nothing could cushion the emotional blow of the exhumation and reburial of over twenty-seven hundred bodies whose gravesites lay in the path of the reservoir. The city paid family members fifteen dollars to move a loved one's body and, in a rare instance of sensitivity to local attitudes, apparently went to great lengths to identify bodies buried in unmarked graves. Nonetheless, one local farmer who was unable to demonstrate a familial connection to those buried on a cemetery at the edge of his land ranted at local undertakers when they came to unearth the bodies, calling them "body snatchers." Most residents were unwilling to do the grisly work of moving their forebears, preferring to pay others to retrieve and transport the bodies to new gravesites outside the taking area.[93] A handful of local residents profited greatly from the disinterring of bodies: West Shokan undertaker Al Schoonmaker unearthed enough bodies to build himself a seven-room house.[94] For most residents, however, the removal of bodies ranked as one of the most painful dislocations of the entire project. Unmoved by the blood-less utilitarian logic that necessitated the exhumations—as reservoir proponent T. Morris Longstreth memorably put it, "the corpses weren't allowed to stand between six million . . . souls and their thirst"—valley residents tried to turn their attention from the past to an uncertain future.[95]

Encountering the City

Much of the conflict between watershed residents and representatives of New York City did not take place in the great outdoors, where board engineers and con-tractors reshaped the landscape without much interference from locals. Instead, it occurred in courtrooms, where homeowners and business owners sought to extract the maximum compensation for property flooded by reservoirs or seized by New York to lay the Catskill Aqueduct. Dissatisfaction with the value of property and damage awards transformed the resentment felt by many watershed residents into full-fledged anger: the city that sundered their communities forced them to fight for every penny. The large volume of cases, the paucity of aggregate data, and the lack of clear criteria by which to interpret their outcomes make it difficult to evaluate the compensation provided to Catskill residents. Attorney Alphonso Clearwater, the project's most prominent opponent, warned his fellow citizens not to trust New York's promises of adequate compensation: "They will say pleasant things to the Catskill mountaineer—and they will get his farm as cheaply as they can."[96] He correctly anticipated the dynamic that prevailed in virtually every com-pensation case: the claimant produced witnesses who attested to a relatively high value for the property or business, while New York countered with its own expert witnesses who invariably pegged the value much lower. Every dollar attorneys could shave off an award represented over a dollar in savings (because of interest on water bonds) to New York.

The scripted quality of expert testimony obscured distinct differences in awards granted to locals. Elwyn Davis, a teenager who lived on his family's farm in West Shokan, recalled these inconsistencies: "Influential people received better awards than small fry, and some commissions gave bigger awards than others."[97] Scrutiny of individual claims reveals additional patterns, especially in business damage cases. Business owners had to demonstrate two things in order to receive substantial awards. First, they had to prove their business was in operation before June 1905, and second, they had to document their income and expenses. These requirements generally posed little problem for owners of stores and large businesses, almost all of whom were men, but proved onerous for many female business owners, especially boardinghouse operators. *gender*

The cases of Delancey Matthews, Martha Young, Rebecca Bonesteel, and Laura Every offer a window into the workings of business damage commissions, a world where legalistic norms and cultural prejudices combined to produce widely disparate outcomes. Rebecca Bonesteel and Martha Young ran boardinghouses out of their West Hurley residences, and their cases, heard by the same business damage commission, seemed to have much in common. Neither woman kept books, and thus had no proof of money received or spent, or of their guests' length of stay. Young testified that she hosted an average of twenty-five boarders a week throughout the summer, and claimed a value of $5,000 for her boarding business; Bonesteel, who tended to twelve to fifteen boarders a week, filed a claim for $2,000. John McGrath, another local boardinghouse operator, testified on their behalf, valuing Young's operation at $3,000 and Bonesteel's at $1,500. However, the two women comported themselves quite differently on the witness stand. Bonesteel, while acknowledging her lack of written records, projected an air of confidence and likely received the commission's sympathy when she clarified that due to her husband's poor health she was the family's sole breadwinner. In contrast, the commissioners appeared to find Young an unsympathetic witness. Exhausted by the specific and detailed questions posed by the city's attorney, who demanded to know the year each family first boarded with her and how many people were in each family, Young became flustered, claiming only that she knew she had each family every year. When the attorney claimed she had contradicted herself concerning the dates guests had stayed with her, Young responded, "I can't tell you anything about the dates, because I didn't keep books." Young also stated that she spent the winters in New York City, where her husband worked. Impressed by Bonesteel's composure and her evident need, but well aware of her lack of documentation, the commission awarded her $600. Likely convinced that Young's husband could support her and unimpressed by her confused testimony, the commission rejected her claim entirely. Thus, Young, who cared for twice as many boarders and actually offered more evidence than Bonesteel about her business—one of her boarders testified on her behalf, and she cited the names of several of her longtime boarders, including a New York City alderman—fell victim to the prejudices of the judicial system.[98]

Laura Every, a young woman who supported herself and her daughter by laundering boarders' clothes, also failed to measure up to commission standards. She filed a claim for $1,000, alleging that reservoir construction provoked an exodus of boarders, destroying her business. Her husband worked on what she called a "tearing-up gang," clearing debris from the reservoir floor. She described a strained marriage in which he only sporadically contributed to the family's upkeep, forcing her to comb his pockets for money when he fell into drunken stupors.[99] Because Every's eligibility for an award hinged on demonstrating that she had begun washing boarders' laundry before the summer of 1905, the city's attorney-for-hire focused almost entirely on her inability to prove when she started her laundry business. After several minutes of persistent questioning, she became flustered, on one occasion stating that she started her business in 1898, only to claim a few minutes later that she began laundering in 1904. When asked to explain this contradiction, she dug herself an even deeper hole, stating, "You twist me so that I don't know how to say it. I can't tell you just the year."[100] The chairman of the commission became annoyed with her, asking, "Have you got any education, Mrs. Every?" After acknowledging her lack of education, Every was forced to display it before the courtroom when she proved unable, at the chairman's instigation, to add six years to 1898. Sensing a lost case, her attorney attributed Every's imprecision to a larger mysterious cultural failing among Catskill residents: "My experience with these people is that they are very loose in their statements as to dates, and cannot fix dates and places. I don't know why, but that seems to be a failing with them."[101] After humiliation came penury; Every did not receive a cent in compensation.

Though her inability to articulate a coherent timeline and the outright rejection of her claim made her case somewhat atypical—of the sixteen cases the commission heard at Every's session, it awarded damages in thirteen of them—it accurately reflected the style and tone of most commission hearings. Rural people found themselves forced to conform to the world of statistics and dates that provided the legal infrastructure for damage cases. Men like Delancey Matthews, a West Shokan general store owner who kept meticulous records of all goods purchased and sold, already operated in this more quantitative world, and received a substantial settlement.[102]

However ambiguous the results of property and business compensation in the final measure, the cultural friction of the commission proceedings intensified residents' distrust of New York City. As early as 1909, only two years after commissions had begun to hear cases, the *Kingston Daily Freeman* adopted a tone of breezy sarcasm when discussing New York City's attempt to persuade an appeals judge to delay granting several real estate awards, observing that the city's "good faith towards Ulster county and its property owners, were once again manifested Saturday before Judge Betts."[103]

Although some valley residents benefited economically in the short term from waterworks construction, many put community cohesion before economic advancement. The tacit agreement to refrain from bidding against one another at auctions where the city sold off homes it had acquired through eminent domain reflected watershed residents' intense desire to reconstitute their communities. In the absence of competing bidders, many families purchased their homes back from New York for a nominal sum, using some of their initial compensation payment to rebuild them on higher ground. City officials soon grew wise to the collusion, opting to burn the homes rather than return them to their owners for fifty dollars. This decision, which was also prompted by the city's desire to accelerate the process of clearing the valley, infuriated residents, who viewed it as wanton destruction of their longtime homes.[104]

New York's political maneuvering also threatened the viability of entire communities. It lobbied Albany for passage of a law that would classify all of the watershed property it acquired for the Catskill system as unimproved land, the least valuable category of land for taxation purposes. Parcels worth hundreds of dollars an acre because they contained fertile land or well-maintained homes and businesses would, overnight, lose virtually all their value, leaving communities like Olive, where the city now owned six thousand acres, with a massive budget deficit. As one Catskill newspaper observed, the legislation would accentuate the physical erasure of valley communities by fiscal means: "The principal point involved and the one that presents the most glaring injustice to the town is that New York city [*sic*] has taken improved lands, village properties, industrial plants and places of business that were assessed accordingly, destroyed all of the improvements and now insists that the land it has made bare should be assessed as unimproved land."[105] The notion that the city could seize valuable land and effectively declare it worthless to reduce its own tax burden, thereby depriving the remaining communities of their largest source of revenue, galled watershed residents. Although New York had a legitimate gripe about the arbitrary nature of assessments in many upstate communities, its extreme response—pursuit of what Ulsterites came to call the "tax-dodging bill"—only further alienated the people of the Esopus Valley.[106] The demise of the tax bill emerged as perhaps the most enduring victory for watershed communities in this era. In towns where New York City acquired substantial chunks of real estate, it sometimes paid as much as three-quarters of real estate taxes, greatly reducing the burden of other taxpayers. The dispute about how to value the city's watershed holdings and infrastructure persisted into the twenty-first century.

Financial disputes between New York City and watershed communities notwithstanding, perceptive observers recognized that money was not the top priority for many Ulster residents. In perhaps the most honest account of the compensation process offered by a city official, Comptroller Metz vividly described the psychology of the displaced farmer: "He, like the rest of us, is a creature of environment. He knows his particular valley. The coming and going of the seasons are to him almost matters of intuitive knowledge. The periods of rain and shine, warmth and cold, and their influence upon growing things in that neighborhood were taught to him in childhood. . . . To give such a man $5,000 for a small farm, although it may have been a generous reward, may not compensate him."[107] The region's children shared farmers' deep attachment to valley rhythms and places. Eleanor Arold recalled that her mother, a young girl at the time her family was forced to move, nursed an enduring resentment of the city.[108] Vera Sickler, a contemporary of Arold's mother who later became town historian in Olive, lamented the physical changes wrought by reservoir construction. She recalled the damage to a spring that ran through her childhood farm, and regretted that New York took no steps to preserve local landmarks such as the covered bridge at Bishop Falls.[109]

Even as the residents of Olive and Hurley struggled to adapt to the changes left in the wake of the construction of the Ashokan Reservoir, the BWS had largely moved on. When the board reflected on its accomplishments, it resorted to a propagandistic recounting of history that completely effaced the resistance of Catskill residents:

> The taking of the Catskill waters has been so accomplished that the immediate regions from which the waters have been diverted have been benefited and beautified, rather than laid waste. Extensive highways of improved character, which are the pride of their localities, have been built by the City, sources of pollution affecting the general health have been cleaned up, and the dams and reservoirs have been so constructed that they enhance the beauties of the landscape. Moreover, all those whose property has been taken or damaged have received just, even liberal, compensation.[110]

CITY OVER COUNTRY

Although this statement appeared in a self-serving report to the mayor, its idealistic depiction of reservoir construction suggests a curiously distant, almost abstract, perspective on rural life, one increasingly shared by urban elites. President Roosevelt's Commission on Country Life, which toured the nation from 1908 to 1909, issued paeans to traditional country values, but it ultimately prioritized production for growing cities over competing rural imperatives. Under the new political paradigm, rural areas existed to serve cities, not as autonomous places whose rhythms and traditions deserved protection.[111]

The passing of the Progressive Era did not vanquish the impulse to impose order on American society and the country's natural resources. On the contrary, it institutionalized a brand of liberalism that emphasized the need for rational planning and the greater good.[112] The construction of the Catskill Mountain water network, which began in the heady days of the Progressive Era and concluded just as the Great Society began to disintegrate, embodied this modern, liberal ethos.

Rural people unable to adjust to this new regime of modernism, who could not produce accounting ledgers and other written records, were at a distinct disadvantage.[113] The development of New York's water system was part of a more general trend of systematization, of large-scale integration of people and resources, which defined twentieth-century America. The parrying between city attorneys, who sought to establish uniform criteria for determining compensation values, and local witnesses, who emphasized the unique qualities of each business or home, embodied the conflicting values of urban and rural America in this era.[114]

New York City's appropriation of the Catskills' natural resources was nothing new, but urban control over the mountains intensified in this era. Control did not always mean aggressive exploitation. In 1885, New York State established the Adirondack and Catskill Forest Preserves. Appalled at the rapid transformation of the Adirondacks by loggers and railroads, Theodore Roosevelt and other urban residents advocated for protection of state-owned lands in the region. Skillful political maneuvering by the region's state representatives in Albany ensured that state-owned lands in three Catskill counties were added to the list of protected areas. The Catskill Forest Preserve was a political afterthought; at thirty-four thousand acres it was only one-twentieth the size of the Adirondack preserve. In 1904, New York State established the Catskill Park, which consisted of all land, both private and public, within a demarcated area, known colloquially as the blue line. The park defined the area that the state would target for future acquisition. The construction of recreational facilities and increased acquisition of land by the state protected the integrity of Catskill landscapes, provided abundant recreational opportunities, and kept commercial development at bay.[115]

Extensive state land ownership in the Catskills safeguarded the quality of the city's water supply. The Ashokan Reservoir lies almost entirely within the borders of the park, as does much of the watershed for the Esopus Creek. The substantial overlap between state-owned lands and the Catskill watershed amounted to a massive environmental subsidy for New York City. Taxpayer-funded land purchases in the Catskill Park enhanced water quality, obviating the construction of an expensive filtration plant to cleanse the city's water supply.[116]

Ironically, dam and reservoir construction, which benefited so much from the presence of the park, transformed the park more than any other form of development. The Ashokan wiped out farms and forests and changed the character of Esopus Creek. After completion of the Ashokan project, activity shifted forty miles north, to the Schoharie region of the Catskills. Construction of the Schoharie Reservoir, which dammed the north-flowing Schoharie Creek, added another 200 MGD to the water supply. The eighteen-mile Shandaken Tunnel reversed the flow of Schoharie's waters, sending them south into the Esopus Creek, where they emptied into the Ashokan Reservoir for delivery to New York.

Nature was a gift, but it was a gift that could be improved through human ingenuity. The damming of free-flowing waters fundamentally changed Catskill streams. Although deforestation, fishing, and farm runoff had undoubtedly altered mountain streams, they retained much of their natural character. With the construction of dams and reservoirs, the streams became "organic machines," subject to the whims of the city that regulated their flow. Urban needs, as much as the rhythm of the seasons, dictated stream ecology, temperature, and flow.[117]

With work on the Schoharie portion of the system under way by 1917, top BWS officials shifted their attention further afield. They began to contemplate the day when the demand for water would exceed the reliable supply from the Croton and Catskill systems. Anticipating the "many years of active agitation" that would be needed to secure permission to develop new watersheds, engineers began to scout new sources.[118] With east-of-Hudson streams still off-limits, the board looked to the west and the Delaware River. A new, unpredictable chapter in New York's water history had begun.

CHAPTER THREE

Drought, Delays, and the Delaware

By the late 1920s, New York City enjoyed a world-class water system. In little more than two decades, the BWS had orchestrated the construction of a massive collection and distribution network that more than tripled the city's available water supply. But few New Yorkers seemed to notice. Construction of the Schoharie Reservoir barely registered in the public consciousness. Businessman Henry Towne, who chaired the Water Supply Committee of the Merchants' Association, lamented, "The public hears and knows nothing about it."[1] The water supply had come to assume its familiar position in the city's psyche: indispensable, but largely forgotten.

Although the building frenzy that swept the city in these years overshadowed completion of the Schoharie, the two developments were linked; the transformation of the urban built environment depended on copious amounts of water supplied by a thoroughly upgraded delivery network. The construction of skyscrapers and a sprawling subway network enabled New York to expand both upward and outward in the early twentieth century. As the working world became increasingly vertical, many residents fled its crowded quarters for the more horizontal and dispersed outer boroughs. Manhattan's skyscrapers depended on an abundant supply of water delivered under high pressure. By tapping Catskill streams and improving infrastructure in Brooklyn and Queens, the city also ensured that new residents of the outer boroughs enjoyed a reliable supply of water. Developers could honestly promise prospective tenants, many of whom had grown accustomed to irregular water availability in the tenements of the Lower East Side, that their new apartments boasted multiple taps and a plentiful supply of water under good pressure. Brooklyn landlords routinely touted running water as one of the notable amenities provided in their apartments. The prospect of improved municipal services that had swayed many outer borough residents to support consolidation with New York had become a reality.[2]

Towne may have rued how quickly New Yorkers came to take a modern water supply network for granted, but he expressed little regret about the results of system expansion. The extension of the water network in the first three decades of the century went largely according to plan. The same could not be said for the period from 1930 to 1950. The board confronted legal and financial obstacles in its quest to further expand the water network by tapping the Delaware River, and the Department of Water Supply struggled to overcome droughts and seemingly inexorable increases in water consumption. The Great Depression and World War II exacerbated these challenges. The romanticism of the early Catskill years gave way to the sober realities of economic contraction and resistance from other states opposed to the city's plans to dam the Delaware.

The defining feature of the 1930s and 1940s was not new construction, but rather the reconstruction of outmoded elements of the water system. New York City eliminated much of its urban and suburban water supply infrastructure, remaking the metropolitan landscape in the process. In the nineteenth century, officials had constructed a dense network of pumping stations and reservoirs within the city to ensure adequate water supply and pressure. By the 1870s, Manhattan was home to four reservoirs, including two in Central Park. As higher-elevation sources and suburban storage reservoirs came online, the urban reservoirs that had proved so vital were no longer needed. Just as expansion of the Croton network rendered the reservoir at Murray Hill obsolete in the 1890s, construction of the Catskill system obviated the need for much of the existing aboveground urban water supply infrastructure. Ironically, it was not until the Depression opened the spigot of federal work relief funds that New York exploited the latent recreational potential of these defunct reservoirs by turning them into parks and swimming pools. The reverberating effects of rural water expansion were not limited to former reservoir sites. Construction of the Catskill system also decisively shaped the development of Long Island. In responding to New Yorkers' ever-increasing demand for water, the city did more than build impressive new dams and reservoirs that most of its citizens would never see. It unleashed forces that produced some of the metropolitan region's most treasured public spaces.

Refashioning old reservoirs proved simpler than controlling nature. The 1940s began and ended in drought, creating temporary but potentially catastrophic water shortages. In response, authorities mounted extensive public relations campaigns urging New Yorkers to watch every drop. When the rains returned, conservation quickly became an afterthought. Rather than meet demand by controlling consumption through water metering, improved leak prevention and repair, and other conservation measures, New York focused almost exclusively on expanding supplies. Neither drought nor an unprecedented economic depression deterred the city from pursuing its imperial prerogatives.

Brooklyn and Long Island

In both the short and long term, New York saved Brooklyn's water supply. The most immediate beneficiary of the introduction of water from the Catskills was Brooklyn, which had been limping along with a patchwork system for decades. Without New York City's assistance, Brooklyn's growth would have been severely stunted by its inability to supply water to its expanding citizenry. Barred by the 1896 Burr Act from tapping new sources in eastern Long Island and short of tax revenue, Brooklyn had no hope of keeping pace with the rapidly escalating demand for water. In the waning days of Brooklyn's independence, its mayor insisted that "no more important matter in relation to Brooklyn has been left to the consideration of the authorities of the new city" than the borough's water supply.[3] New York used its resources to repair leaking reservoirs and drill new wells, enabling Brooklyn to muddle through until the delivery of water from the Catskills. The new East River bridges that connected the two boroughs received most of the credit for Brooklyn's rapid growth, but it was the water that flowed through City Tunnel No. 1, beneath the river, that ultimately secured the borough's future. By 1930, Brooklyn was home to two and a half million residents, far more than Manhattan; its population had more than doubled in the three decades since it became part of New York City.[4]

When mountain water finally arrived in 1917, Brooklyn residents reacted with glee. After decades of inadequate water pressure and genuine fears of shortages, the borough could finally count on a reliable supply. Of the 140 million gallons consumed by Brooklyn's citizens in a typical day, all but 30 million gallons came from upland sources. With water flowing from the Catskills, officials could abandon many sources on Long Island that had become increasingly polluted in recent decades.[5] The city also dismantled Brooklyn's forty pumping stations, selling the land to builders and developers.[6] The effects on Long Island were less innocuous. Island landowners claimed that Brooklyn's reduced diversions were turning previously dry parcels (whose runoff had been pumped away to Brooklyn) into swampland, and threatened to sue New York. In a sign of the intimate connections between urban needs and rural landscapes, these property owners relied on Brooklyn's thirst to maintain their lands. The hydrological ripple effect was bizarre, but real: tapping the distant watersheds of the Catskill Mountains meant flooding portions of the sandy plains of Long Island.[7]

The most enduring consequence of decommissioning many of Brooklyn's wells and reservoirs was the creation of a recreational oasis for hundreds of thousands of New Yorkers. Before Robert Moses overhauled the landscape of New York City in his role as chairman of the Triborough Bridge Authority, parks commissioner, and several other municipal posts, he transformed Long Island. As president of the Long Island State Park Commission, Moses created the most elaborate system of parks and parkways anywhere in the country. From the packed shores of Jones Beach to the remote reaches of Montauk on the eastern tip of the island, he opened up an island that had been largely off-limits to city residents. Without the water-supply lands made available by the construction of the Catskill water system, he likely would have failed.[8]

In his quest to make Long Island accessible to New York City residents, Moses confronted a pair of seemingly intractable obstacles: Long Islanders fiercely opposed the designs of outsiders, and he had only a limited pool of state money to acquire and develop parks and parkways. His first inkling of a possible solution came in 1922 while gazing out the window on a train from New York City to Babylon, a village on the south shore of the island where he rented a summer cottage. He noticed that dense patches of forest predominated between many of the villages. Learning that the woods belonged to the city's Department of Water Supply, Moses began to eagerly explore them, discovering a remarkably untouched assortment of ponds, lakes, and reservoirs rimmed by massive old trees. Brooklyn had acquired these properties in the nineteenth century to build and protect its water sources, almost all of which were located in Nassau County, directly to its east. Arrayed like beads across the western portion of the island, the city's thirty-five hundred acres of watershed property included prized parcels such as the Hempstead Reservoir. The second-largest body of fresh water on Long Island, Hempstead also included hundreds of acres of buffer lands ideal for horseback riding, hiking, and other forms of recreation.[9]

Brooklyn's watershed lands, Moses realized, could help him remove the obstacles that made the island inaccessible to urban residents eager to sample its charms. In the early 1920s the automobile became a staple of middle-class life. Tens of thousands of New Yorkers hit the road on summer weekends, eager to picnic and swim on Long Island's shores. The recalcitrance of private property owners and regulations barring nonresidents from using town facilities rendered the island a tantalizing mirage: a world of sand, surf, and cool breezes was within reach, but remained outside the grasp of hordes of city dwellers. Just beyond the view of a parade of frustrated drivers lay "land, a vastness of land, overflowing with the fruits they sought. Land that didn't have to be condemned. Land that was owned not by the governments of Long Island, which viewed the masses of New York City as foreigners, but by the government of New York City."[10]

With help from Governor Al Smith, his political benefactor, Moses secured easements on more than half of Brooklyn's former water supply holdings in 1925. The natural abundance and strategic location of these twenty-two hundred acres represented "the most valuable single addition to the Long Island State Park and Parkway System."[11] Moses carved a series of parks from these former watershed lands, which stretched from just east of the city line clear across most of

Nassau County. New Yorkers who had struggled to find accessible parkland could now frolic on the water slides at Valley Stream State Park or go horseback riding under the oaks at Hempstead.

The value of these watershed lands went well beyond their recreational assets and natural features. They enabled Moses to realize his grand vision: knitting the island together with a series of parkways that made the shoreline accessible to tens of thousands of urban dwellers. The first crucial link in this transportation network, the Southern State Parkway, provided access to the watershed parks and, in connection with other new roads, to Jones Beach, the jewel of the park system. Without the watershed lands, the Southern State Parkway would never have been built. While much of the parkway was cobbled together from lands donated by or purchased from private property owners, the western section, the stretch closest to Brooklyn, consisted largely of former watershed lands. With these watershed lands in hand, Moses generated momentum for the parkway, which encouraged landowners to sell him the parcels he needed to make it a reality.[12]

New York City's watershed lands and its recreational needs were intimately bound up with the suburban development of Long Island. A stretch of the Southern State Parkway, one of the island's most traveled roads and a critical conduit for commuters to the city, wound around a portion of Brooklyn's former Hempstead Reservoir, which became the focal point of Hempstead State Park. When Governor Smith came to Long Island in November 1927 to mark the completion of the parkway's first section, he traveled to the park, where a tablet commemorating the event was set in the retaining wall of the Hempstead Reservoir.[13] Although Moses recognized that the new parkways would ease travel for the island's residents, he consistently emphasized the benefits of parkways and parks for urban dwellers. He rejected proposals for parkway routes suggested by Long Island residents, insisting that the parkways must connect to the city's existing road network.[14]

It seemed only fitting that Brooklyn's former watershed lands provided the key to help unlock Long Island's forests and beaches for millions of city dwellers. New York invested heavily in Brooklyn's water supply in the nineteen years between city consolidation and the introduction of Catskill water. As New Yorkers streamed out to Long Island to swim in its waters and inhale the fresh air of its woods, municipal leaders must have marveled at how quickly and abundantly they had reaped the karmic dividends of saving Brooklyn's citizens from thirst.

Parks in the City

Well before Robert Moses envisioned the recreational potential of Brooklyn's watershed lands, municipal officials had established strong links between New

York's water supply infrastructure and its parks. By the turn of the century, virtually every reservoir within city limits was surrounded by parkland. The remote (at least at the time of their construction) location of the reservoirs, the need to provide a protective buffer around them, and the belief that they represented a desirable park amenity bound New York's parks and reservoirs together.[15] While similar logic guided officials in Boston and Philadelphia, waterworks expansion did more to shape New York's parks than those of any other American city.[16]

In the 1890s, as New Yorkers fiercely debated the fate of the Croton Reservoir at Murray Hill, Brooklyn officials began to develop the area surrounding Ridgewood Reservoir, on the border between Brooklyn and Queens. Brooklyn acquired land around the reservoir for park purposes, but the most significant improvements, including a rustic bridge, construction of a music stand, and reclamation of a swamp for a flower garden, occurred after consolidation with New York. Its expansive views of Brooklyn and the Atlantic Ocean, and the elaborate flower garden—deemed "one of the most unique and superb seen in any of the parks hereabouts"—made Highland Park became one of New York's most beloved small parks.[17] Well aware of the paucity of such desirable parcels, officials expanded the park by purchasing a private estate and developing a third parcel owned by the Department of Water Supply. Highland's twenty-eight tennis courts and extensive ball fields became a major recreational destination for thousands of outerborough residents. By the 1920s, ballplayers mingled with children tending farm gardens and the thousands attending summer band concerts.[18]

The connection between recreation and the water system dated back even further in Manhattan. Central Park had been home to two distributing reservoirs since the 1860s, when the upper reservoir was built to supplement the original reservoir, which lay just to the south, between 80th and 86th Streets. With the arrival of water from the Catskills in 1917, the lower reservoir was disconnected from the water distribution system. By 1934, Robert Moses had added New York City parks commissioner to his growing list of titles. He saw an opportunity to put discarded water system property to good use once again. One of his first initiatives as parks commissioner was to clear the makeshift homes destitute New Yorkers had erected on the site of the filled-in reservoir, in order to build a vast green area, which came to be known as the Great Lawn. Moses hewed closely to a plan developed by the New York chapter of the American Society of Landscape Architects that envisioned an oval-shaped green space to be surrounded by a handful of playgrounds. The fifteen-acre expanse of grass was originally largely off-limits to the public, but by the 1950s the Great Lawn's ball fields had become Manhattan's recreational hub.[19] In the 1980s, it hosted some of the largest concerts in the city's history and a massive antinuclear rally. Those interested in more solitary pursuits ambled and jogged around the upper reservoir, whose 106 acres stretch from 86th to 96th Streets. The original focus of water system recreation, however, lay well to the north, in the High Bridge area of Washington Heights on the Harlem River.[20]

Since being built in 1848 to carry Croton water over the Harlem River, High Bridge had attracted visitors from around the city. In the warm months, New Yorkers boarded boats in lower Manhattan for the journey up the East and Harlem Rivers to the bridge. By the 1880s, High Bridge had become one of the city's most popular day trips. *Harper's Weekly* rhapsodized, "Nothing can be pleasanter, for those who have only a short time at their disposal, than an afternoon trip to High Bridge, where the scenery is delightful, and where one can enjoy the sight of the great structure over which rushes the supply of water for New York, take a walk over the high banks, or sit on shaded benches to watch the rowers on Harlem River."[21]

High Bridge's appeal lay in the intermingling of the natural and the manmade. The steep banks of the Harlem River inhibited development, ensuring the preservation of a long but narrow stretch of woods. In the 1870s, the city acquired

Fig. 7. High Bridge, Harlem River. At top, the elegant arches of the High Bridge. Pedestrians enjoyed ambling on the bridge on weekends. Note the water tower in the background of the main image. (*Harper's Weekly*, August 22, 1885)

additional land on high ground above the bridge, where it built a reservoir and a striking Romanesque Revival granite water tower. Constructed to improve water pressure for high-elevation customers, these additions considerably enhanced the accessibility of the area. Visitors could promenade around the reservoir and gaze down on the river valley below, or pay a small fee to ascend an iron spiral staircase to the top of the water tower, which offered a glimpse of the Hudson and commanding views of northern Manhattan.[22]

By 1910, High Bridge anchored a stretch of recreational attractions in northern Manhattan. In the 1890s, the city purchased additional land to create Highbridge Park, which soon spanned more than twenty city blocks.[23] The park's natural cliffs and large rock outcroppings made it one of New York's wildest destinations. Visitors with more genteel tastes drifted south to the Speedway, where the wealthy raced their horses beside the Harlem River. North of the park, thousands of New Yorkers caught trolleys to Fort George, where a popular amusement park opened in 1895. When a fire destroyed the amusement park in 1914, the city purchased the grounds, further extending the northern boundary of Highbridge Park.[24] By the 1920s, Highbridge was one of Manhattan's largest parks.

Even before the Great Depression bore down on New York, Highbridge and most other parks had lost their luster. Equipped with meager recreational facilities and too small to accommodate the swelling population, the parks ranked among the most neglected of municipal assets. After consolidation in 1898, the city established separate parks departments for each borough, leading to a predictably incoherent recreation policy. Mayor Fiorello La Guardia's decision to consolidate responsibility for all parks under Robert Moses in 1934 inaugurated a golden age of recreation in New York.

Beginning in the late 1980s, historians began to reevaluate Moses's legacy. Complicating Robert Caro's portrait of a tyrannical and insensitive public works dictator, these scholars argued that Moses, for all his faults, immeasurably improved the public life and landscape of the city.[25] The most enthusiastic praise came from scholars who reexamined his performance as parks commissioner. They argued that the familiar portrayal of Moses as the architect of highway construction and slum clearance who destroyed the homes and communities of tens of thousands of New Yorkers represented only one dimension of his complex accomplishments. Before he became a master of slash-and-burn urban planning in the 1950s, he had taken a softer approach to reshaping the city. By acquiring a wide array of idle or underused city-owned lands—a procedure he dubbed "reassignment"—and wrangling huge appropriations from the Works Progress Administration (WPA), Moses greatly expanded New York's recreational resources. In doing so, he retooled waterworks infrastructure to meet the growing needs of the city's densest neighborhoods.[26]

The water supply system directly and indirectly shaped the development of one of the core components of Moses's recreational agenda: the construction of a network of massive (some were larger than Olympic size) swimming pools throughout New York. Many cities tapped WPA funds to build swimming pools, but none did so with the fervor of New York.[27] By the summer of 1936, Moses's ability to secure a disproportionate share of WPA money for New York ensured that citizens of every borough enjoyed access to large outdoor swimming pools.[28] In typically declamatory language, Moses explained why a city with such extensive waterfront access launched a massive swimming pool construction program: "It is one of the tragedies of New York life, and a monument to past indifference, waste, selfishness and stupid planning, that the magnificent natural boundary waters of the City have been in a large measure destroyed for recreational purposes by haphazard industrial and commercial developments, and by pollution through sewage, trade and other waste."[29]

As its population grew, New York deposited increasing volumes of raw sewage in its rivers and harbors. Another important feature of the region's complex hydrological commons was the geographical separation of drinking water and sewage. The ability to draw on distant supply sources led New York to treat its waters as an enormous sink for wastes. Had the city opted to draw its water from the Hudson (all such proposals called for locating the intake far upstream of New York, where the water was less polluted, and for aggressive treatment, including filtration), it likely would have substantially upgraded its sewage treatment facilities out of fear that tides would push severely contaminated water upstream. In the 1930s, the federal government launched a campaign to purify the country's sewage-laden rivers and harbors, and the city used an infusion of WPA funds to initiate a much-needed expansion and modernization of its sewage treatment plants. Moses strategically deployed these resources to clean up the waters adjacent to the city's most attractive beaches, on Coney Island and the Rockaways. Residents of the Bronx and Manhattan could on occasion travel to these beaches, but Moses sought to provide them with a much more convenient alternative: neighborhood pools.[30]

Highbridge Pool, one of eleven large outdoor pools opened in the summer of 1936, typified Moses's approach to aquatic recreation. Like most of the new pools, it was located in a congested neighborhood—Washington Heights—and built on city-owned property. Impressive, but less architecturally elaborate than some of the other new pools, Highbridge featured an attractive brick bathhouse with a capacity of almost five thousand, and large swimming, wading, and diving pools. The pool's grandeur derived largely from its unique setting. Swimmers gazed out at the stately water tower hovering just outside the pool grounds. After swimming, they could amble over to the landscaped promenade near the tower's base, from which they enjoyed stunning views of the High Bridge and the Harlem River valley.

Highbridge Pool's connection to the water system went beyond its proximity to the bridge and the water tower; the pool was carved from the foundations of the Highbridge Reservoir, which was taken out of service when the New Croton Aqueduct and a new pumping station came online in the 1890s.[31] It sat idle until 1934, when the DWS transferred the reservoir, the water tower, and the surrounding grounds to the Parks Department. The refashioning of the reservoir into Highbridge Pool took Moses's principle of adaptive reuse of existing public property to new heights. At Highbridge, Moses did not simply build a pool on unused city-owned land; he took advantage of the physical form of the property to more efficiently construct a new recreational resource.[32]

Virtually overnight, Highbridge became a treasured neighborhood resource and one of the most popular pools in the city. Undeterred by admission fees and long lines, Washington Height's working- and middle-class Jews and Irish flocked to the pool. Washington Heights resident Herbert Levenson wrote Moses to thank him for providing his daughter with a recreational resource usually reserved for those with greater means: "I want to tell you how we folks who do not own cars and do not necessarily use the roadways appreciate the swimming pools you have built for us. I cannot send my youngster away for the summer but even so as long as she can go down to the High Bridge pool for the afternoon she is delighted."[33] Another neighborhood resident, Jane Rush, complained to Mayor La Guardia about the poor access for baby carriages at Highbridge, but echoed Levenson's enthusiasm: "That pool was such a God send to us that I don't know what to tell you in appreciation."[34] By converting the reservoir into a swimming pool, the Parks Department turned the customary griping about the relentless architectural metamorphosis of New York into delighted screams of glee from thousands of the city's children. Changes in the countryside, in conjunction with federal funds and active city planning, had turned a vestigial wart on New York's landscape into a shining new symbol of recreational and civic health.

A similar pattern emerged in the northeast Bronx, where the Williamsbridge Reservoir had sat idle since the construction of the Catskill system. Built in 1887 as a distribution reservoir for water conveyed from the Bronx and Byram Rivers—modest sources that supplemented the Croton system—Williamsbridge was located in a densely populated neighborhood. The *New York Times* enthusiastically endorsed Moses's plan to convert the nineteen-acre reservoir site into a recreational complex: "The man who found tongues in trees, books in the running brooks and sermons in stones had nothing on Commissioner MOSES. He discovers recreational facilities in the most unexpected places; for him even the toads have 'precious jewels' in their heads."[35] Moses retained the reservoir's oval shape, but trimmed the thirty-seven-foot-high embankment walls and raised the basin's floor twelve feet. Pedestrians who strolled along the inner portion of the embankment walls looked down on a two-level recreation facility complete with a wading pool, tennis courts, playgrounds, and ball fields. Workers recycled both the reservoir site and the materials of the reservoir itself; a field house was built largely from stone taken from the reservoir, "a most economical and attractive material for the purpose," noted the Parks Department.[36] A four-season resource that offered everything from dance competitions to ice-skating, Williamsbridge Oval soon became a fixture of the neighborhood.[37]

If Moses's feats did not quite reach the biblical heights suggested by the *Times*, he succeeded in transforming unproductive infrastructure into highly desirable cultural and recreational amenities. Even as the board struggled to find the resources to construct a new water supply system during the lean years of the 1930s and 1940s, WPA laborers busily reconstructed the old one, converting defunct reservoirs into tennis courts, pools, and playgrounds. As Moses had demonstrated on Long Island, the water system could play a significant role in expanding recreational opportunities for New Yorkers.

The churning quality of the built environment—the quick turnover of its buildings, blocks, and neighborhoods—has long been one of New York's defining

Fig. 8. Girls' Dance Festival, Williamsbridge Oval, June 1945. This photo depicts one of the many recreational and cultural functions held on the site of the former Williamsbridge Reservoir. (New York City Parks Photo Archive)

characteristics. As early as 1836, former mayor Philip Hone noted the fleeting quality of urban geography: "The old downtown burgomasters who have fixed to one spot all their lives, will be seen during the next summer in flocks, marching reluctantly north to pitch their tents in places which, in their time, were orchards, cornfields, or morasses."[38] As historian Max Page observed in his study of Manhattan's transformation in the first half of the twentieth century, New Yorkers generally viewed the whir of change with a mixture of pride and regret. Eager innovators, they nonetheless rued the disappearance of favorite places. Page and most other historians identified the pulsing energy of capitalism as the main force behind the city's frequent face-lifts. In fact, the imperatives of both private capital and public need reshaped the city.[39]

The most notable aspect of the redevelopment of former water supply properties was the reverberating quality of regional landscape change. These reverberations were a product of the continuous development of the water system outside the city. Construction of the Croton system in the nineteenth century scattered water infrastructure throughout Manhattan. Extension of the Croton network and construction of the Catskill system turned these reservoirs and pumping stations into relics. The engineers who labored on the Ashokan and other Catskill projects may have felt far removed from the city and missed its attractions, but in building a new water system in the countryside, they laid the foundation for the remaking of many critical urban spaces. The dramatic transformation of rural landscapes is an obvious legacy of New York's thirst for water. Less obvious, but equally transformative, was the reshaping of suburban and urban landscapes in the wake of New York's incursions into the Catskills.[40]

Going to the Delaware

The politics of water changed in the 1920s and 1930s. The search for additional supplies for the expanding metropolis was no longer primarily a battle between rural and urban New Yorkers. External forces such as resistance by other states, the fluctuations of the national economy, and an increasingly forceful federal government challenged New York's ability to deliver that most basic and precious of resources, water.

These external problems were rooted in an internal reality: New Yorkers' thirst for water appeared limitless.[41] Fueled by explosive growth—from 1900 to 1930 the city's population more than doubled to nearly seven million—and sharply higher per capita consumption due to the Catskill system's increased water pressure, the diffusion of indoor plumbing, widespread leaks, and profligate water use, overall consumption increased dramatically.[42] The completion of the Schoharie Reservoir

in 1926 would provide a reprieve until 1935, but because of the long lead time before the board could reasonably expect to receive permission to build additional
reservoirs, it began planning for a new supply as early as 1920.[43] By 1922, the new
chief engineer, Thaddeus Merriman, had charted an ambitious course: New York
would consider tapping tributaries of the Delaware River in Delaware County,
northwest of the Ashokan Reservoir.

From a political perspective, Merriman's proposal to pursue Delaware sources
looked almost as audacious as George Washington's crossing of the river. Few
American rivers play as important a role in defining political geography as the
Delaware. The main stem of the river is formed by the confluence of two Catskill
Mountain streams, the East and West Branches of the Delaware River, which con-
verge in the town of Hancock, on the Pennsylvania border. The Delaware divides
New York and Pennsylvania for the length of its eighty-two-mile course to the New
Jersey state line. It remains a border stream for its entire journey south to the Dela-
ware Bay, dividing New Jersey from Pennsylvania, and New Jersey from Delaware.

In addition to supplying water to many cities, including Trenton and Philadelphia, the Delaware fostered the region's economic growth in the early twentieth
century. The inland anchor of New Jersey's thriving recreation industry, the river
attracted thousands of fishermen, boaters, and outdoor enthusiasts who packed
the hotels and resorts of the Delaware Water Gap and the surrounding area.[44]
Further downstream it powered the boilers of manufacturers and chemical plants
that made southeastern Pennsylvania an industrial hub.

The proposal to tap an interstate stream like the Delaware had significant legal
implications. Instead of simply securing state permission to build new dams and
reservoirs, New York City would need state officials to negotiate (or, in the worst-
case scenario, litigate) on its behalf. Bringing Delaware water to New York City
would require either an interstate compact with New Jersey and Pennsylvania to
divvy up water rights, or a decision by the United States Supreme Court granting
the city the right to divert water from the river.

Why venture into the roiling waters of interstate law? Nature did not dictate
the decision; many sources located entirely within New York State, including
the Hudson River, a network of streams east of the Hudson, and other Catskill
Mountain streams, were still available. But Merriman, who was one of the first
BWS employees and served as Chief Engineer J. Waldo Smith's longtime right-
hand man, shared the board's well-established preference for upland water supplies. Like the Esopus Creek, Delaware water came from a sparsely populated
mountain watershed. Therefore, it required no filtration and little or no pumping
before delivery to consumers. The quality, abundance, and relative affordability of
Delaware water made it an obvious choice for New York's next supply.[45]

Merriman's determination to exploit Delaware sources appeared to reflect a realization that his legacy would rest more on his ability to resolve the thorny issue

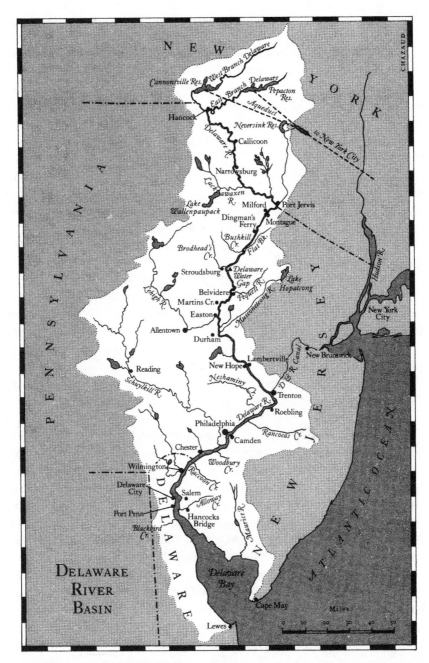

Fig. 9. Delaware River basin. The Delaware River is the border between several states in the Mid-Atlantic region. Their competing claims to the river generated two lawsuits decided by the U.S. Supreme Court. (Jacques Chazaud Design; the map originally appeared in Bruce Stutz, *Natural Lives, Modern Times, People and Places of the Delaware River*, 1992.)

of interstate water sources than on the design of any particular dam or aqueduct.[46] Confident that the Delaware held enough water to satisfy the needs of the entire region, he initially envisioned a unified project under which a single aqueduct stretching from New York State to Philadelphia would provide water to all three states. Such a plan, he insisted, would make use of untapped waters, reduce flooding, and, through a series of well-timed water releases from the city's reservoirs into the Delaware, ensure more vigorous flows in summer months. The challenge lay in convincing New Jersey and Pennsylvania that New York City's proposed diversions from the Delaware would in fact yield such benefits. Merriman insisted that "the Delaware River situation is a clear-cut instance of the importance and need for interstate cooperation and agreement." Taking his cue from the recently signed Colorado River Compact, which allocated the river's waters between upper and lower basin states, he called for the creation of an interstate commission to allocate Delaware waters.[47] In 1925, less than three years after outlining his vision for the river, he got his wish when the governors of the three states appointed commissioners to forge an interstate compact to divide the Delaware.

Although the benefits of interstate cooperation seemed clear to Merriman, they looked far murkier to many in New Jersey and Pennsylvania. Each state appointed commissioners to craft a deal that would be presented to their respective legislatures. The most vehement opposition to a compact came from New Jersey's capital city of Trenton, which drew its water supply from the Delaware. Trenton feared that the daily withdrawal of hundreds of millions of gallons of water by New York City would reduce river flow in the summer and diminish the Delaware's capacity to dilute contaminants. While Trenton's concerns were firmly rooted in the language of the proposed agreement, its objections went deeper. The city's counsel, Merritt Lane, argued that the unpredictable effects of such a large diversion made a compact an inappropriate legal instrument because New Jersey would lack the power to repeal its provisions. As Lane observed, "It is fair to assume that when you interfere with the works of nature to the extent that you intend under this compact there is going to be some result not perhaps anticipated."[48] In contesting New York's proposed diversion, he suggested that the inflexibility of a compact agreement was at odds with the inherent unpredictability of nature.

From New Jersey's perspective, maintaining the status quo appeared more sensible than relying on Merriman's assurances. Nature may have been unpredictable, but in Lane's mind interstate water law was not. Confident that the Supreme Court would find against New York because common law traditionally proscribed transfers across watersheds, he urged state legislators to reject any compact. Because New Jersey did not anticipate increasing its withdrawals of Delaware water until 1980, there was simply no compelling reason to sign a deal that provided so little benefit to the Garden State. Pennsylvania also had doubts about the wisdom of agreeing to a compact. By 1927, Merriman's optimistic vision of interstate cooperation appeared increasingly quixotic.[49]

Stymied by this resistance, the BWS decided to draw on the Delaware without the permission of neighboring states. With Pennsylvania's legislature out of session until 1929, prospects for ratifying the compact and delivering a new supply by the target date of 1935 appeared remote. To delay, argued the board, was "fraught with risk."[50] In March of 1928, the state legislature approved New York City's plan to divert tributaries of the Delaware River.

The unilateral decision to tap the Delaware represented an enormous gamble by the board. The plan, which called for diverting six hundred million gallons a day and included a reservoir release schedule designed to increase river flows in the dry summer months, mirrored the terms of the compact that the states had failed to ratify. From the board's perspective, New York was asking for no more than had already been agreed to by water commissioners from the three states. Nevertheless, Merriman and his colleagues recognized that engineers and lawyers did not always see eye to eye. New Jersey or Pennsylvania, having rejected the compact, might very well seek an injunction to prevent New York from initiating construction. If New York lost the case, its position would become perilous. By the mid-1930s, water consumption would nearly equal the available supply. With the Delaware foreclosed, the city would have to obtain permission to develop an in-state water source. Even if it secured speedy state authorization for a new project, New York would likely experience several years of water shortages until additional supplies came online. Despite the enormous political, economic, and social risks, the board refused to abandon its insistence on developing additional mountain supplies.

With cooperation no longer an option, the BWS opted for persuasion, mounting a carefully choreographed campaign to convince other states that New York would prevail in court.[51] To buttress its position, the board solicited opinions from prominent legal authorities, including former solicitor general and Democratic presidential candidate John W. Davis. Davis argued that the "line of precedent" originated in a 1907 case, *Kansas v. Colorado*, in which the Supreme Court first articulated the principle of equitable apportionment. Under equitable apportionment, states had a right to divert water from an interstate stream if their diversions did not cause serious injury to other riparian states.[52] The willingness of the water commissioners of New Jersey and Pennsylvania to agree to the compact convinced Davis that any injury to these states would be minor. Moving from conventional legal reasoning to the rhetorical flourish of a closing argument, he suggested that the city's economic importance inoculated it against an adverse decision: "The Court would doubtless take judicial notice that New York City is the metropolis and business center of the United States, and that the entire country is vitally interested in the regular functioning of the financial and commercial system of which it is the heart."[53]

Because many observers perceived a more nebulous legal landscape than Davis, the city's bravado did not have the desired effect. As Davis recognized, the critical question concerned the likely injuries that New Jersey and Pennsylvania would sustain if New York dammed Delaware tributaries.[54] Under the riparian doctrine, the basis of water law in the eastern states, courts settled water disputes on the principle of "reasonable use." Courts often disagreed sharply about what constituted reasonable use (the term had no statutory definition and thus was subject to widely varying interpretations), but a consensus had emerged that the rights of some water users could be curtailed so as not to minimize the enjoyment of a stream by others holding water rights. Many state courts had ruled that diverting water outside its watershed, as called for under New York's Delaware plan, represented an unreasonable use.[55] As hydrologist and environmental planner Maynard Hufschmidt astutely observed, this uncertain legal context made New York City "particularly anxious to obtain by interstate agreement what might possibly be forbidden to it by Court decree."[56] The city would soon learn the strength of its legal position, because in May of 1929 New Jersey filed suit with the Supreme Court seeking a permanent injunction against the Delaware project.

The involvement of multiple parties and the likelihood of an extended hearing replete with technical testimony led the justices to assign the case to a special master, Memphis lawyer Charles Burch, who was directed to take evidence and submit a recommendation for a decree. The justices reserved the right to approve, reject, or modify Burch's recommendations.[57] When the Court granted Pennsylvania's petition to join the case as an intervenor, the stage was set for a lengthy hearing. Neither a plaintiff nor a defendant, Pennsylvania nonetheless had the right to file testimony and call witnesses. It did not formally oppose New York's bid to divert Delaware waters because it anticipated that it would eventually need to tap upstream portions of the river to supply Philadelphia's growing population.[58] Eager to preserve these future rights, but skeptical of the efficacy of New York City's proposals for releasing water from its reservoirs to maintain river flow in the summer, Pennsylvania, in its status as an intervenor, was akin to a back-seat driver whose views actually held some weight.

Despite the array of parties and the technical complexity of the issues involved—testimony ran to over seventy-five hundred pages, and the parties filed over four hundred exhibits—the case boiled down to two fundamental questions: Was New York City legally entitled to divert water from an interstate stream, and would the diversion of 600 MGD seriously injure New Jersey?

To answer the first question, Burch relied almost exclusively on interstate water law. Slightly rewording the Court's ruling in *Kansas v. Colorado*, the foundational case of interstate water law, he defined the task before him: "The problem here is to adjust the dispute so as to secure to New York the benefit of the use of the tributaries of the Delaware for municipal water supply and at the same time not deprive New Jersey of the benefits of a flowing stream."[59] After rejecting New Jersey's rather extreme claim that the common law of riparian rights prohibited any diminution in river flow by an upstream user, Burch dismissed the state's other overarching argument—that interstate common law prohibited diversions from one watershed to another—noting that the Supreme Court had explicitly endorsed such trans-watershed diversions. Burch found further support for his position in the laws and practices of the three states. Each state had permitted diversions across watersheds; in fact, New Jersey diverted substantial amounts of water from the Delaware River into the Delaware and Raritan Canal.[60] In his view, to prohibit New York from tapping the Delaware would deprive it of its right to make reasonable use of such waters.

Having established New York's right to divert the Delaware, Burch then considered whether the withdrawals would seriously injure the welfare of New Jersey and its citizens. Reconciling the highly divergent opinions of scores of expert witnesses on subjects as varied as navigation, oystering, recreation, and waste dilution proved a formidable task. Burch asked the parties to refrain from filing exhibits not easily understood by a layperson, and consistently sought to winnow expert testimony by highlighting what he deemed the most critical determinants of injury.

In assessing the probable effects of the diversion on the river, he emphasized the crucial issue of timing. What mattered was not the overall reduction in stream flow due to New York's withdrawals, but whether such reductions were likely to occur in summer, when dry weather considerably reduced stream flow.[61] The city repeatedly stressed that the reservoir releases called for under its proposal would actually increase summer flows. Intently focused on the real-world implications of New York's release plan, Burch criticized the proposal for its potential lack of sensitivity to actual conditions, observing, "There may be a freshet [sudden overflow of a stream] or flood in the river at Port Jervis when no more water is needed, and yet under the New York plan, water would be released as an addition to this flood or freshet."[62] His decree recommended the adoption of an alternative release plan proposed by Pennsylvania that modified the release protocol to give more weight to actual downstream conditions.

After sifting through mounds of testimony, Burch granted New York City most of what it sought. He pared the city's permitted diversion from 600 MGD to 440 MGD to reflect the likelihood that siphoning off over half a million gallons of water a day from the river's tributaries would seriously impair New Jersey's use of the Delaware. Although he determined that the plan would do little damage to

New Jersey's industrial base, agricultural production, or municipal water supplies, Burch concluded that permitting such a large diversion threatened New Jersey's recreational and oyster industries. Diminished river flow would likely make swimming and boating on the Delaware less appealing, damaging the "reputation of the river as a place of recreational resort."[63] By reducing the flow of fresh water into the river, the larger diversion would also increase salinities where the Delaware River emptied into the bay. Higher salinities would encourage the growth of oyster predators, and a less productive oyster industry endangered the economic livelihood of thousands of New Jersey residents.[64]

In addition to calling for a reduced volume of diversion and adoption of Pennsylvania's reservoir release plan, the decree recommended that New York City construct a sewage treatment plant for the city of Port Jervis, New York, to improve the quality of the Delaware River before it flowed into New Jersey. In Burch's estimation, his revised plan would substantially reduce damages incurred by New Jersey, rendering New York's withdrawals a reasonable use.[65]

When the Supreme Court affirmed the special master's decree in May 1931, New York City finally exhaled—it would get its Delaware water after all. Although the decree reduced the city's allotment of water from Delaware tributaries by more than 25 percent, New York declared victory. Chief Engineer Merriman prized the flexibility that came with obtaining the water through a court ruling rather than an interstate agreement, which likely would have imposed many more restrictions on the city's use of the river. He concluded that "440 m.g.d. without a compact is worth more than 600 m.g.d. with one. Hence it seems to me that we have won a glorious victory."[66]

New Jersey grumbled, but the engineering community held Burch in high regard. Even some of New Jersey's own expert witnesses, such as prominent civil engineer Abel Wolman, praised his efforts: "I am completely amazed at the thorough-going review and grasp of the subject which the report indicates."[67] Wolman's praise substantiates Merriman's predictably enthusiastic endorsement of Burch's report: "His findings on the law are excellent, while on the facts they are as good as any one had a right to expect."[68]

A modest man who modeled himself on his mentor, J. Waldo Smith, Merriman omitted any mention of his own critical role in influencing the outcome of the case. However, Burch's report revealed the extraordinary power of individuals to shape legal outcomes. Burch cited Merriman's testimony as justification for his own conclusions on several crucial aspects of the case, including river flows and the superiority of Delaware water over other potential sources.[69] After sifting through thousands of pages of testimony, which "consumed all of my days and nights for the past few months," Burch disregarded much of it, focusing on those he deemed the most relevant witnesses.[70] Despite the report's thoroughly professional tone, Burch's clear admiration for Merriman shines through: "He has given a most careful, intelligent and intensive study to the problems of this case."[71]

Vanquishing New Jersey was easier than overcoming the restrictions imposed by the Great Depression and World War II. The difficulty of financing the Delaware system and wartime manpower and materials shortages severely hampered construction, at times bringing it to a complete halt. The first new reservoir, the Rondout, was not put into service until 1951. The Pepacton Reservoir, the project's main collecting basin, first impounded water in 1954, and did not fill completely until 1956, roughly two decades behind schedule.[72]

Water Conservation

Delays in constructing the Delaware system revived a long-running debate about water conservation, and in particular about the wisdom of installing water meters in private homes and apartment buildings. The discussions revealed a great deal about both the power dynamics of municipal government and New York City's self-identity. The reluctance to implement aggressive conservation measures ensured the continuation of water supply expansion, a policy with far-reaching environmental and social implications.

The economic crisis significantly delayed construction of dams on the Delaware. New York struggled to sell the bonds required to finance the Delaware system, slowing construction progress. If the steady annual increases in per capita and overall water usage of the 1920s had continued, the city might well have run out of water. But as the economy contracted, so did water consumption. From 1930 to 1932, per capita consumption decreased by almost 10 percent, returning to 1920 levels. Although population growth led to higher overall usage by the end of the decade, annual consumption increases remained relatively modest.[73]

Nonetheless, officials at the Department of Water Supply began to worry. Even small annual increases in consumption compounded over time, and by 1940 department officials began to question New York's ability to withstand an extended drought.[74] Focused on moving the Delaware project ahead, their counterparts at the BWS largely ignored concerns about the narrowing margin between supply and demand. When the board did acknowledge the other side of the equation, it invariably portrayed increasing demand as an inevitable, and desirable, fact of modern existence: "Improvements in our mode of living tend towards a more liberal use of water and a liberal use of wholesome water is a recognized index of public health."[75] Mayor La Guardia also equated high water usage with a vigorous body politic, boasting, "We use 990 million gallons of water a day, which speaks well for the cleanliness of our people."[76]

What the mayor failed to mention was that much of this water was not put to productive use. Critics lamented the wintertime ritual of leaving faucets running to prevent frozen pipes, but waste attributable to such practices paled in comparison to the leaks that riddled the system. A 1940 WPA study exposed a porous city; an extensive survey of thousands of buildings uncovered more than four hundred thousand leaks. Investigators estimated that waste due to faulty plumbing fixtures and aging water mains totaled approximately 200 MGD, an astounding one-fifth of daily deliveries.[77]

The WPA study confirmed what many engineers and journalists had been saying for years: New York's water management and delivery policies were utterly retrograde. Commissioner of Water Supply Maurice Davidson contrasted the high quality of New York's water with its abysmal administration: "It is a sad commentary . . . that while it stands preeminent in its source of water supply and distribution, it has failed to keep abreast of modern control and business practices, with the result that it is the outstanding great backward city from the standpoint of purveying water to its consumers, still functioning under an obsolete system of flat rates instituted in the year 1857."[78] Under the flat rate system, the city charged residential consumers based on the dimensions of their property (as a result, water rates were frequently dubbed "frontage rates"), not on their actual consumption. Rates increased with the size of the building and the number of water fixtures it contained.

Building size proved an exceptionally poor proxy for water consumption. Differences in family size and the prevalence of leaks were only two of the factors throwing off the equation. The city operated a split system in which it metered all business and industrial usage, charging by actual consumption, but applied the unreliable frontage system to residential properties. The failure to align the rate structure with actual consumption resulted in a massive cross-subsidy to residents and landlords. Although they consumed three-quarters of New York's water, they paid only 55 percent of the tab.[79] Critics called for the installation of meters in residential buildings. Residents and landlords continued to waste water, they argued, because higher use did not result in higher bills. Employing a favorite analogy of reformers, Commissioner Davidson compared water use to other utilities: "It has well been said that no person dreams of letting a gas stove burn when it is not in use, or of leaving electric lights turned on when they are not needed, but a dripping faucet may be neglected for months."[80]

A mundane and largely invisible device of modern life, water meters had long been a source of controversy in New York. By the turn of the century, many American cities, large and small, had substantially or entirely metered water systems. In city after city, metering reduced consumption, often by as much as 20 or 30 percent.[81] Beginning with John Freeman's landmark investigation in 1900, virtually every report on New York's water system recommended extending metering to include residential customers. Nonetheless, by 1911, metering of consumption had emerged as a political third rail.

In 1910, the prospects for expanded metering appeared bright. Mayor William Gaynor recruited Edward Bemis, who had overseen a universal metering initiative in Cleveland, to tackle the problem of water waste. Soon after arriving in New York, Bemis realized that high consumption rates were at odds with the city's residential and commercial structure: "In the first place, there are comparatively speaking, no lawns here, and, second, the iron and steel business concerns of Western cities give way here to lofts, housing clothing manufacturers and similar users of waters." New York's manufacturers required relatively modest amounts of water. The same could be said of its apartment dwellers, who lacked the lawns and shrubs common in less densely populated cities. Bemis had struck on New York's dirty little secret: its per capita consumption, instead of exceeding or matching that of other large American cities, should actually have been lower.[82]

Bemis's hiring signaled Mayor Gaynor's intention to modernize city government. That his arrival coincided with the onset of a fairly severe drought boded well for an extension of meters; the political and meteorological stars appeared to be aligned. Then, under the slogan "Water should be as free as air," opponents of metering counterattacked. Arguing that meters would compel New Yorkers to curb even justifiable use of water, owners of apartment buildings mounted a vigorous campaign against metering.[83] Mayor Gaynor did an about-face. He fired Bemis and adopted the alarmist language of the real estate elite: "If heads of houses had to pay by meter, they would be uneasy when their wives and children took baths. The result would be discomfort and uncleanliness. . . . I believe in getting an inexhaustible source of water supply and letting every one use all the water he or she wants for washing and bathing and domestic purposes."[84] The city hired extra personnel to detect leaks in water mains and to conduct house-to-house inspections. It also launched a publicity campaign urging residents to conserve water.[85] These short-term measures cut water use by 20 percent, allowing New York to ride out the drought, but they did not reduce water consumption over the medium and long term. Metering, for the time being at least, was dead.

Renewed calls for expanded metering emerged periodically, usually coinciding with a drought or proposed expansion of the water supply system. Thwarted by city politics and the myth that "'spiteful tenants' would let the water run just to increase their 'hated landlord's' bill," proponents of metering looked to outsiders to force the city's hand.[86] With Albany reluctant to crack down on New York, advocates shifted their gaze to Washington. They hoped that the federal government would make access to the Delaware conditional on implementation of universal metering.

New Jersey advanced this argument before the Supreme Court, contending that New York should conserve existing supplies before diverting water from interstate streams. Chief Engineer Merriman countered that the expense of universal metering outweighed the potential water savings. The city also presented evidence that its per capita water use was in line with that of other cities. As Bemis had observed two decades previously, this claim was highly misleading. New York's dearth of residential lawns and its slight industrial base should have resulted in lower per capita consumption than in industrial cities such as Buffalo and Pittsburgh, even after factoring in consumption by tourists and commuters. Burch largely dismissed other opinions on water conservation, focusing on Merriman's testimony.[87] However, he did seize on the observation of one New Jersey witness, who estimated that universal metering would delay the need for the Delaware project by five years. Burch claimed that because it would take eight to ten years to complete even the first stage of the Delaware project, New York was "not acting prematurely in attempting to arrange for an additional supply of water." In other words, given the scale of the Delaware project, metering was largely beside the point.[88]

This argument—that the scale of New York's thirst dwarfed the likely effects of universal metering—proved difficult to refute. Meter advocates observed that curbing consumption would delay the start date of future water expansion projects, saving the city money in the short term. They also emphasized that universal metering would provide New York with a new supply of water for only a tenth the cost of building new dams and reservoirs.[89] However, because the city repaid water bonds with revenue derived from water rates, which were quite low, these arguments had little political resonance. Curbing consumption would not mean more money in taxpayers' pockets.

Conservation advocates did not aggressively question the unspoken premise of urban democratic imperialism—that the needs of the city (the many) always outweighed the preservation of the countryside (the few). This utilitarian vision was in turn based on the assumption that the city would continue to grow, if not forever, at least into the foreseeable future. The board, in large part because its sole mission was to expand the water supply, refused to contemplate the day when population growth would cease or slow dramatically. When growth finally ebbed, the need for incremental water supplies would evaporate, or at least significantly abate. Lowering per capita (and therefore overall) water consumption through universal metering and implementing a reasonable rate schedule would hasten the arrival of this day, a fact that Burch failed to appreciate. Given the size of New York City's population, universal metering could conceivably reduce consumption by hundreds of millions of gallons a day—the savings would mount as the city's population increased—obviating the construction of an enormous reservoir or two, thereby preserving the economy and way of life of another section of the Catskills.

Only a handful of observers connected the reluctance to fix drips and leaks with the physical transformation of Catskill landscapes. Even those who did, like George Waring, the celebrated reformer who oversaw a highly successful cleanup of New York City's streets in the 1890s, generally emphasized the financial flaws, not the environmental effects, of an expansionist policy: "This is the old, old story. Only one way seems to suggest itself to our water-works engineer for meeting an impending shortage. This is, to go farther and farther afield and to draw on remote water-sheds, at whatever cost."[90] In 1927, the *New York Evening Post* drew the connection between municipal policy and the transformation of the countryside, noting that constructing new water supplies resulted in the loss of "beauty spots" in the Catskills.[91]

Dealing with Drought

A combination of delays on the Delaware project and increasing consumption left New York vulnerable to water shortages during prolonged droughts. Lacking any alternative, officials launched massive public relations campaigns urging residents to reduce their water consumption. Despite two major droughts in the 1940s, municipal leaders remained adamantly opposed to universal metering.

When a drought hit the region in 1940, officials formulated a menu of policy responses they would rely on for decades to come. In what one magazine called "the largest program of water conservation ever attempted," the city worked closely with the private sector to spread the gospel of saving water. From dawn to bedtime, in private and public settings, New Yorkers were reminded of the drought. Hotel guests found notes in their rooms asking them to conserve the city's water; "Stop That Leak" posters covered subway walls; radio stations issued frequent warnings not to waste water; and theaters throughout New York played newsreels featuring shots of Mayor La Guardia engaged in earnest discussions with his water supply commissioner. In addition to beefing up leak detection crews, the city called on schoolchildren to serve as its eyes and ears. Civics teachers instilled the virtues of water thrift, urging their students to report any leaks and monitor their parents' water use. Instead of relying on the impartiality of water meters, the city depended on the guilty consciences and watchful eyes of millions of New Yorkers. What was widely regarded as the country's most technologically sophisticated water system took a resolutely low-tech approach to surviving the drought.[92]

City officials and newspaper editors recognized the need to balance high-minded calls for conservation with a dash of whimsy. A DWS spokeswoman took to the airwaves in January 1940 to recite a poem reminding listeners that:

> In course of time a dripping tap
> Can drain a river off the map,
> And Monday's water poured unheeded
> On Tuesday may be sorely needed.[93]

A month later, cartoonist Rube Goldberg joined the fun. His water shortage suggestions included reducing the size of bathtubs and taking a Saturday night vacuum in lieu of a bath.[94] New Yorkers may not have heeded Goldberg's call to wear newspaper underwear to reduce the amount of laundry they generated, but they did markedly curb their consumption. Within four months, water use decreased nearly 10 percent. When the rains finally arrived, New Yorkers returned to business as usual.

The arrival of war the following year posed new challenges for the water network. Although work on the Delaware system slowed considerably, making conservation all the more crucial, municipal officials ignored the rapid deterioration of water infrastructure within the city. Concerned about possible German contamination of the water supply, the city expanded its watershed police force. Meanwhile, water consumption skyrocketed, increasing by 25 percent in only five years, a dramatic spike that could not be explained by the booming economy or population growth.[95] City officials attributed the increase in part to the difficulty of obtaining repair materials during the war. A 1947 investigation identified leaking plumbing fixtures in unmetered buildings as the primary source of waste, echoing the results of the WPA study. With completion of the first stage of the Delaware project still several years away, water officials sounded an ominous note: "For the period 1948 to 1956, the City is faced with a very serious problem."[96]

The prediction was soon realized: the city confronted the worst drought in living memory. Beginning in the summer of 1949, municipal officials called on citizens to reduce their water consumption. New Yorkers initially ignored the warnings of impending shortages, consuming a record amount of water that summer. But when fall failed to bring substantial rainfall, the alarms became increasingly dire. Intimately familiar with their roles, all the cast members in the water conservation drama played them to the hilt. Officials relied on a combination of new regulations and a massive publicity campaign to curb water usage. They launched "Dry Fridays," a voluntary program in which New Yorkers were asked to make special efforts to conserve water on Fridays by taking shorter showers, or none at all.[97] The DWS also banned lawn watering, and imposed new regulations intended to reduce water used to power air conditioners, which accounted for an astounding one-sixth of consumption during summer months.[98]

Beseeched by their government, residents again slashed their water consumption dramatically. Likely eager for an excuse, many men willingly refrained from

Fig. 10. Rube Goldberg, "Water Shortage Suggestions," originally appeared in *New York Sun* during the 1940 drought. Goldberg offered some unorthodox tips to help New Yorkers conserve water. (*New York Sun*, February 19, 1940)

shaving on Fridays. Water Commissioner Stephen Carney trumpeted the civic virtue of facial hair: "A Friday beard," he observed, "is a beard of honor."[99] Other citizens had tired of this ad hoc approach to water conservation. In a letter to the editor, Susanne Bedell urged her neighbors to consider the roots of the present water crisis: "We Americans are a curious tribe. Give us a good crisis and we shall rise to the occasion ingeniously, bravely and always successfully. Sometimes, however, we should ask how the crisis arose and point to the source."[100] *Herald Tribune* editors agreed, dubbing the belated banning of water-guzzling air conditioners "a costly mystery."[101]

Such criticism did little to alter the familiar contours of the water conservation debate. Fearing an increase in their bills, property owners continued to challenge the efficacy of meters, arguing that they would only measure waste, not reduce it.[102] City leaders showed no sign of ending their long-standing opposition to metering and other policies intended to curb consumption. Mayor William O'Dwyer sounded almost apologetic when he was forced to impose a series of conservation measures in October 1949: "I have been reluctant to go to the public with any appeal for the curtailment of water usage. But it is impossible to defer any longer taking some definite measures to cut down consumption."[103] In the midst of a severe drought, New York's leaders focused as much on increasing the supply of water as on curbing demand. The short-term solution to boosting supplies was the construction of a temporary water intake on the Hudson River at Chelsea, eighty miles upstream from New York. The mayor also appealed to BWS president Irving Huie, asking him if everything was being done to expedite the Delaware project. "Could men work on Saturdays?" inquired O'Dwyer.[104]

The most controversial aspect of the city's plan to weather the drought was to change the weather. Despite the sharp dip in consumption, reservoir storage levels continued to decrease in early 1950. Desperation began to set in. Cardinal Spellman's call for priests and parishioners to offer special prayers for rain yielded meager results. Without significant rainfall, New York could expect to run out of water by the summer. Once again, a drought inspired poetry. In March, the New York Philanthropic League lamented, "We've got the Hudson and the Atlantic / But the water shortage drives us frantic."[105] Having tried virtually everything else, the city asked Wallace Howell, a Harvard meteorologist, to make it rain.[106]

Howell's hiring proved perhaps the most infamous, and certainly one of the most comical, incidents in the history of New York's water supply system. Equipped with police department planes, Howell proposed bombarding promising cloud formations with dry ice in the hope of coaxing rain from them. Ironically, inclement weather ruined Howell's first attempt to "seed" Catskill clouds. Due to overcast skies, he was delayed in flying out of Brooklyn. By the time Howell and his crew reached the Catskills, skies had cleared, ending any possibility of rainmaking for the day.[107]

The city insisted on proceeding with cloud seeding despite concerns among watershed residents that the experiments would exacerbate spring floods. When Howell's efforts coincided with a mid-April snowstorm, the region's residents fumed. Cursing the city's "damn shenanigans with dry ice," one small-town selectman noted that he had already removed the plows from the town's trucks. "I was caught," he seethed, "with my plows off. We in Newtown are indignant and lay without hesitation the blame at the door of the Mayor's office."[108] Having failed to secure an injunction banning the practice, Catskill residents filed damage suits, alleging that rainmaking caused the severe flooding that ravaged the region that spring. City attorneys successfully claimed that the rain and snow that had wreaked such damage in the Catskills and replenished reservoirs were nature's doing, not Howell's. A newspaper headline neatly summarized New York's apostasy: "City Now Skeptic on Rain-Making."[109]

Consigning the rain-seeding experiment to the realm of entertaining but meaningless anecdote would be a mistake. At one level, the incident offers another potent example of New York's imperial attitude toward its watersheds. Just as the effects of global warming will fall hardest on those countries least responsible for causing it, so did the possible consequences of rain seeding inflict the most harm on the dairy farmers and hotel owners least able to fall back on alternative forms of income. Looked at from a different perspective, the incident reveals a city willing to place its faith in an unproven, short-term technology—rain seeding—rather than the established, long-term solution of metering. New York's leaders, ever focused on the supply side of the equation, did not heed these lessons. Instead, they pointed to the drought as Exhibit A in their bid to launch yet another round of waterworks construction. With the drought scare behind it, New York looked forward to building monuments much grander than a nondescript residential water meter.

The Limits of Power

This confidence reflected both a continuing belief in the mission of water supply expansion and New York's self-conception as the world's most powerful metropolis. In the rush to look forward and finally complete the long-delayed Delaware project, New York's water chieftains ignored important warning signs. Despite the looming threat of a water shortage, officials emphasized the need to increase supplies, largely ignoring the potential benefits of conservation measures. Even when the Great Depression and World War II disrupted the board's timelines, delaying construction by more than fifteen years, municipal leaders shunned calls for a different approach. The logic of perpetual growth left no room for a more conservative approach. Municipal officials viewed the mid-century droughts as a reason to accelerate reservoir construction, rather than a cue from nature to adopt a more holistic approach to managing the city's water supply.

But the ideology of growth only flowered in the right political soil. The split governing structure of New York City's water system strongly favored continued expansion of supplies. In flush times, New York's decision to divide responsibility for its water supply between two agencies, with the board building the system that it then turned over to the Department of Water Supply to operate, worked well. The BWS could concentrate on system design and expansion and leave the details of daily operations to an established city agency. When external forces retarded system expansion, bifurcation of responsibilities led to suspect policy decisions. Buttressed by the resistance of property owners and the expansionist vision of the board, New York's mayors and legislative leaders ignored calls by engineers to pursue aggressive long-term water conservation measures. A single agency charged with both system construction and daily operations likely would have prevailed on municipal leaders to adopt a more balanced approach that relied on a combination of supply expansion and demand reduction.[110]

The city's expansionist agenda led it to pursue a policy of selective cooperation. It engaged in serious negotiations with New Jersey and Pennsylvania to divide the Delaware, but its inability to transcend its own parochial interests thwarted a deal. Similarly, New Yorkers curbed their water consumption during droughts, but returned to their profligate ways when the rains returned.[111] Compromise proved a useful tool for managing crises, not an underlying principle of the city's water management strategy. Despite a stupefying lack of foresight that forced the world's most prosperous city to rely on a combination of incessant pleading (with both its own citizens and God), the unproven technique of rainmaking, and well-timed rains to weather the drought of 1950, New York remained committed to a unilateral program of dam and reservoir construction. Underpinning this commitment to supply expansion was the assumption that the population would continue to increase substantially. But the tremendous growth of the 1920s and 1930s slowed in the 1940s. Nevertheless, the board looked to build additional reservoirs in the Catskills.[112]

With new construction slowed in the countryside, the most significant changes to the water system occurred in and around New York City. Through the force of his personality and vision, Robert Moses reengineered urban and suburban infrastructure to create some of the most significant recreational spaces in the metropolitan area. In an era characterized by waiting—for rain, for the economy to rebound, for the war to end—Moses, his engineering and architectural colleagues, and legions of WPA laborers tapped the water system for its recreational potential. In the decades to come, however, New York resumed its frenzy of building activity in the countryside, bringing wrenching social and ecological changes to another section of the Catskills.

CHAPTER FOUR

Back to the Supreme Court

The 1950s and 1960s marked a critical transition period for the city and its water system. New York supplanted London and Paris to become the center of the Western intellectual and artistic world.[1] Mayor Robert Wagner Jr., who presided over City Hall from 1954 to 1965, expanded an array of housing and health programs, cementing Mayor Fiorello La Guardia's legacy of generous social benefits for New Yorkers.[2] He also supported redevelopment projects masterminded by Robert Moses that eviscerated the neighborhoods of many of these citizens. The technocratic, top-down ethos of urban renewal bore a striking resemblance to the Board of Water Supply's style of operation. Not surprisingly, Moses was one of the board's staunchest supporters.[3] The BWS, unmoved by political resistance and stagnating population growth, called for the construction of additional mountain reservoirs.

The drive for further expansion of Catskill supplies embodied the city's muscular self-confidence. But New York was not immune to the trends that were rapidly transforming America's physical and sociopolitical landscape. World War II and the Cold War decisively shifted the nation's political center of gravity to Washington, D.C. New Yorkers joined the stampede to the suburbs; for the first time since the Dutch arrived, the city's population declined.[4]

Suburban growth was inextricably linked to continued expansion of New York's water supply. Rapid residential and commercial development in Westchester County placed enormous pressure on local water sources. Cities and towns throughout the county increasingly resorted to purchasing water from New York City to meet the escalating demand from businesses and residents. Just as the development of the Catskill system had transformed the urban landscape in the 1930s and '40s, access to New York's water played a key role in shaping Westchester's built and natural environments in the postwar period.[5]

Of course, the most direct impact of supply expansion occurred in the Catskills, where thousands of residents lost their homes and businesses to make way for another round of reservoir and aqueduct construction. By the mid-1950s, laborers had completed three new reservoirs—Rondout, Neversink, and Pepacton. Rondout and Neversink collected the waters of intrastate streams; the Pepacton was formed by damming the East Branch of the Delaware River, which merged with the West Branch to form the main stem of the Delaware. Tunnels connected the Neversink and Pepacton Reservoirs to the Rondout, where water from all three reservoirs intermingled before entering the Delaware Aqueduct, an eighty-four-mile conduit that carried the runoff of the western Catskills to New York. The BWS had overcome years of false starts, economic depression, and legal battles to bequeath New York "its own private miracle: a daily flood of nearly two billion gallons of cool, pure mountain water to quench the thirst and supply the ever increasing domestic and industrial demands of millions of people living on the ocean's edge."[6]

The familiar rhetoric glorifying supply expansion obscured a new reality: a wide range of dissenters began to challenge New York City's water policies. Civic organizations, the press, and outside experts—all of whom had generally supported the city's approach to water acquisition and management—proposed alternative policies and challenged the board's integrity. These critics raised fundamental questions about New York's water strategy. Was the Hudson River really too polluted to serve as an alternative water supply source? Was the city's erratic pattern of releasing water from the Pepacton Reservoir killing fish and altering the weather in the Catskills? Why did New York's water conservation program lag so far behind other cities?

The board and its allies in municipal government struck back, proudly reiterating their commitment to an abundant supply of pure mountain water at nominal rates. Why sully the country's premier public water supply with the industrial refuse of the Hudson? Why did New York receive so little recognition for the superlative new trout fishery that its releases from the Pepacton Reservoir spawned on the East Branch of the Delaware River? What other city in America provided its citizens—rich and poor alike—with pristine water for only pennies a day?

The critics of New York's water regime confronted many of the same obstacles facing opponents of urban renewal. In each case, the city's unwillingness to seriously consider opposing viewpoints and its penchant for grand construction projects infuriated those who envisioned an alternative future. Neighborhood activist and writer Jane Jacobs accused officials of overlooking the complex interconnections that bound a neighborhood together and ensured the prosperity and security of its residents—of misunderstanding its ecology.[7] BWS foes questioned

New York's decision to build additional reservoirs, its resistance to conservation, and its unwillingness to acknowledge the adverse ecological effects of its reservoir release practices. As the 1960s wound down, opponents of urban renewal began to win some victories, most notably the shelving of the Lower Manhattan Expressway. But the BWS and municipal officials made only modest concessions to their detractors. In the realm of water policy, the needs of urban residents trumped rural desires and environmental considerations.[8]

The city's single-minded pursuit of its own water needs represented the last gasp of urbanism. By the 1970s, cities would no longer wield outsize political influence. Financial difficulties and racial tensions sapped the energy of cities throughout America. The rise of the environmental movement and the resulting state and federal environmental regulations constrained New York's management of its water network. Completion of the Delaware water system represented both the realization of a vision that dated back to the 1830s and the end of an era, but its consequences would reverberate for decades to come.[9]

The Board Flexes Its Muscles: Damning the Hudson

New Yorkers greeted 1950 warily; the drought still lingered. Despite sharply reduced consumption due to an ambitious water conservation campaign, the water supply remained precarious. Terrified by the possibility that water shortages could stall the state's economic engine, Governor Thomas Dewey demanded that New York expand its supply. The only viable short-term option was to pump water from the Hudson River into existing aqueducts. In January 1950, the BWS received state approval to temporarily draw water from the Hudson, a project expected to increase supplies by 10 percent. The decision to tap the Hudson opened up a Pandora's box of questions about the water system and its future.[10]

The construction of a Hudson River pumping plant threatened to undermine both the board's traditional role and the city's reliance on upland sources. The board left the day-to-day management of the water system to the Department of Water Supply; its sole mission was to identify, collect, and deliver additional supplies to the city. Developing a temporary supply to help New York weather the drought blurred these lines. More important, constructing the Hudson plant legitimized a water source that the BWS had consistently disparaged as polluted and unfit for human consumption.[11] Forced to swallow its principles because of the drought, the BWS emphasized the extraordinary and temporary nature of the decision to tap the Hudson. The Hudson supply would be chemically treated and thoroughly mixed with pure upland waters. In addition, it would sit in reservoirs for several months, long enough to eliminate most impurities.[12]

The agreement to build the emergency plant could not have come at a more politically inopportune time for the BWS. It was in the final stages of preparing an application to the state to develop the third phase of the Delaware project, a dam on the river's West Branch, which would create the Cannonsville Reservoir. This massive undertaking required a significant political and financial commitment: it would cost an estimated $140 million to dam the river, construct the reservoir, and build a forty-four-mile tunnel to carry Cannonsville's water to the Rondout Reservoir for delivery to New York via the Delaware Aqueduct. Building Cannonsville would also mean returning to the Supreme Court to obtain permission to increase the city's daily allotment of water from the Delaware. All those opposed to further development of Delaware River sources could now point to the Hudson as a plausible alternative to Cannonsville. Less than a month after the state approved the Hudson for use as an emergency supply, the Citizens Budget Commission (CBC), a prominent New York City civic organization, presented the board with a plan to tap the river as a permanent water source.[13]

The CBC's proposal, known as the Beck Plan after its author, engineer Lawrence Beck, directly challenged the board's fundamental approach to expanding the water supply. Upland supplies, argued the CBC, were both increasingly impure and inadequate; the BWS later acknowledged that additional supplies from Cannonsville would only prove sufficient until 1980 if water consumption continued to escalate.[14] In sharp contrast, the enormous volume of water flowing down the Hudson provided an almost "inexhaustible reserve." Moreover, tapping the Hudson by constructing a barrier dam would cost significantly less than continuing to develop mountain water sources.[15]

Although the debate between the CBC and the board focused on the validity of these claims and on Beck's qualifications, the source of New York's next water supply had important symbolic dimensions, as both sides were quick to recognize. From the board's perspective, tapping the Hudson meant abandoning its long-standing principle of obtaining the best water available. Such apostasy was unthinkable to engineers who viewed the construction of dams, reservoirs, and aqueducts to convey pure water from the Catskills as the hallmark of their tight-knit organization. The excellence of the city's water was also a critical component of New Yorkers' self-identity. One high-level BWS engineer emphasized the importance of high-quality water to New Yorkers' civic pride: "The use of the water of the Hudson River for municipal supplies is offensive to the people of the city of New York who are accustomed to drinking water widely and justly regarded as being the best in the world."[16]

The CBC vigorously challenged what it considered the board's faulty logic. It acknowledged that the use of supposedly pristine mountain water "is claimed to be of great psychological importance to the City's character and inhabitants," but insisted that increasing pollution in Catskill watersheds and the insufficiency of such sources for the long term made the construction of another Delaware reservoir a dangerous proposition. By clinging to the "myth" of pure, abundant upland supplies, the board's approach exposed New Yorkers to great risk of a water shortage while new reservoirs were under construction and during droughts.[17]

These dueling visions dominated New York's water politics for the next several years. State officials shared the board's preference for mountain supplies. After perfunctory public hearings at which Catskill residents voiced their opposition to building yet another reservoir and Beck advocated for tapping the Hudson, the state approved the Cannonsville project in the summer of 1950.[18] The real political battle occurred at the municipal level, where a constellation of forces was gathering to challenge the board's Delaware project. Mayor Vincent Impelliterri, forced to choose between the starkly divergent recommendations of the board and the CBC, opted for the elixir of compromise: he commissioned a study of the water system by outside experts.[19] The easing of the drought gave New York time to carefully consider its water policy. In November 1950, with the study commission beginning to take shape, the *Times* endorsed the study, calling for a "disinterested, expert judgment on what the city's course should be."[20]

The study commission, known as the Engineering Panel on Water Supply, offered the first fresh assessment of New York's water supply in fifty years. It included several national experts, most notably Abel Wolman, a professor of engineering at Johns Hopkins University and arguably the most prominent sanitary engineer in the country. Its members pored through internal BWS memos, conducted independent investigations of potential water sources, and reread the board's foundational documents, in particular the 1903 Burr Commission report. Wolman's handwritten comments on these documents indicated that he did not accept many of the board's fundamental assumptions. In the margins of a 1948 BWS report that questioned the suitability of the Hudson as a permanent water source, Wolman scribbled, "Why would an emergency supply be safe, while its permanent use is unsafe? A problem in semantics."[21] He also vigorously challenged the board's claim that it had conducted thorough studies of the Hudson: "Where are all of these studies?" he wanted to know after months of sifting through reports.[22]

The centerpiece of the panel's report was the recommendation that New York draw its next permanent water supply from the Hudson River. It called for an independent study of the most efficient means of tapping the Hudson, and suggested that the city proceed with a preliminary design for the Hudson system.[23] Wolman had testified on behalf of New Jersey in the 1931 Delaware River case, and briefly served as chair of an interdepartmental board of water control established by Mayor La Guardia in 1944. Thoroughly familiar with the details of New York

Fig. 11. Abel Wolman in the early 1950s. (Ferdinand Hamburger Archives, Sheridan Libraries, Johns Hopkins University)

City's water system and no political naïf, he fully appreciated the significance of selecting an industrial river like the Hudson over mountain sources, and insisted that the report highlight the Hudson proposal above all other recommendations.[24]

The panel justified its recommendation on the basis of practicality and distant historical precedent. The utilitarian argument for selecting the Hudson rested on two primary factors: cost and quality. Tapping the river near Hyde Park (substantially north of the location recommended by Beck) would cost only a third of what it would cost to build Cannonsville, even after accounting for the filtration plant required to purify Hudson water. Locating the intake further upstream, where pollution problems were less severe, ensured that "by ordinary standard filtration processes Hudson River water can be made to equal, if not better, the quality of water now supplied to New York City."[25] The panel also emphasized that the notion of drawing on the Hudson had a long historical lineage. Almost a half-century earlier, the vaunted Burr Commission report proposed pumping the Hudson after tapping the Esopus, and concluded that an intake near Hyde Park, when coupled with modern filtration techniques, would yield an "entirely satisfactory" supply.[26]

The recommendation to tap the Hudson represented far more than a simple disagreement over the most appropriate water source for the city; it was a revolutionary call for a new political ecology of water in New York. The panel contended that piecemeal expansion of the supply (New York could gradually increase the volume of water it pumped from the Hudson over time), coupled with vigorous conservation efforts, would provide a safe, abundant supply of water. There was no need to dislocate thousands of rural New Yorkers and spend hundreds of millions of dollars to build yet another reservoir in the Catskills. Tapping the Hudson would instill a conservation mind-set that hitherto had prevailed only in times of drought. As panel member Louis Howson explained, "One of the most important advantages of the Hudson supply is that the higher operating costs will focus attention on possible means of accomplishing savings, whereas with upland storage, in which water wastes over the dam 9 years out of 10, it is impracticable to develop any enthusiasm for expenditures for waste restriction except in those years in which shortages actually occur and then it is too late."[27] In a pumped water system like the Hudson, less consumption meant less pumping, reducing operating costs. Because the Delaware system relied on gravity to convey water from its source to consumers, operational costs were negligible, but the capital outlay required to construct the dams, reservoirs, and aqueducts far exceeded the cost of building pumping and filtration plants on the Hudson.

The report's strong language underscored the political and philosophical chasm that divided the panelists and the board. Wolman and his colleagues explicitly staked their claim to the professional high ground in the report narrative, directly challenging the authority and competence of the BWS. While the panel "approached its task without any preconceptions as to what it would find," it claimed that the same could not be said for previous discussions about the suitability of the Hudson, which had been "lengthy, heated, and not always completely objective."[28]

The engineers' first challenge was simply getting their report released. The panel completed its report in July 1951, and then spent the next five months fighting for its publication. Threatened as never before, the board struck back. BWS president Irving Huie orchestrated a campaign to undermine the panel's report. He condemned its conclusions as false and misleading, and sought to delay or ban its release. The long delay between the report's completion and its formal publication gave the board ample time to prepare a rebuttal report, which it distributed to members of the subcommittee whose stamp of approval was required to advance the panel's proposals.[29]

By Thanksgiving, the press had learned of the conflict and began to agitate for the release of the panel's report. On November 26, a headline in the *World-Telegram and Sun* asked, "Where's the Water Report?" Four days later, the paper declared, "There was as much delay around it as though it dealt with our daily output of A-bombs, or our plans for bombing Moscow."[30] When the report was finally distributed to officials and the press in early December, it was accompanied by the board's dueling analysis. The backroom battle had now become a public feud pitting a distinguished group of engineers against a powerful municipal agency.

The dispute centered almost entirely on the quality of the river's water. The board repeatedly played its trump card—the public perception that the Hudson was a severely polluted river.[31] The BWS reprised the denunciations it had made of the Hudson in condemning the Beck Plan, despite the panel's recommendation that New York tap the Hudson several miles farther upstream, where water quality was far superior. Panel members challenged the board's portrayal of the river, emphasizing that state studies of the Hudson "indicate a water which bears no resemblance to that so disparagingly described by the BWS."[32] Wolman and his colleagues mounted a vigorous public relations offensive to counter BWS claims that the Hudson posed a threat to the quality of New York's water supply. Wolman stressed that the city relied on the same chemicals used by other water systems to ensure the safety of its supply, an indication that New York's water quality was not vastly superior to that of other major cities.[33] At a hearing on the panel's report, he proposed a blindfolded taste test between filtered water drawn from the Hudson and Catskill water, a suggestion that even Hudson proponents did not pursue.[34]

why?

As the panel's report made its way up the political food chain, support for it eroded. When a municipal board rejected the call for an independent analysis of the river, it provided Mayor Impelliterri the political cover he needed to officially kill consideration of the Hudson in November 1952. After more than two years of study and discussion, the plan to tap the Hudson was officially dead.

Had, as the *Tribune* claimed, the decision about New York's next water source "been settled on a more or less emotional basis—that it is better to continue with the Delaware, no matter what complications may arise or how costly"?[35] Perhaps. Certainly, the BWS had successfully exploited fears about the "psychological palatability" of the Hudson.[36] Other factors also sunk the proposal to tap the Hudson. Mayor Impelliterri, a weak leader, was loath to challenge the advice of his water officials. Despite its strong support for the type of conservation measures recommended by the panel, the Department of Water Supply shared the board's fierce opposition to the Hudson, dubbing the river "a highly hazardous source."[37] Given this united front within his administration, Impelliterri's decision to reject the Hudson project came as little surprise.

In the end, a combination of the board's skillful manipulation of the political system and various structural factors helped Cannonsville prevail over the Hudson. In every respect, Cannonsville was the easy choice for municipal leaders. They would not have to allay concerns about the quality of Hudson water.

Nor would they have to raise taxes or water rates. The city would finance the project with long-term bonds, making the cost to the individual water consumer negligible. In contrast, pumping the Hudson would increase operating costs, likely resulting in substantial rate increases. New York would have to secure Supreme Court permission to build Cannonsville, but the board felt confident that the Court would again rule in the city's favor. When it came to water, expertise took a back seat to politics.

The negative repercussions of waterworks construction on the ecology of the Catskills and the region's residents had no influence over the selection of New York's next water source.[38] Opting for Cannonsville would impose tremendous costs on a few thousand residents of the Catskills, but required no sacrifice on the part of city dwellers. Noel Perrin, a Dartmouth professor who vacationed in the Catskills, neatly expressed this disjunction: "Distant villages and valleys are ravaged by the blades of bulldozers so that careless and wasteful New Yorkers can let their leaky taps run."[39] As Wolman and his colleagues observed, tapping mountain streams provided little incentive to curb water consumption because operating costs did not increase as New Yorkers used more water. Spreading the cost of the project over decades and among millions of ratepayers ensured that New York's water rates, which were well below the national average, would increase only marginally. Abandoned by Albany and almost entirely disregarded by New York City, Catskill residents hoped Washington would come to their rescue.

The Delaware River Case: Round Two

The legal landscape had changed since New York City first sought to tap Delaware waters. The 1931 Supreme Court ruling clearly established the city's right to divert water from the river. The question this time around was whether New Jersey and Pennsylvania could convince the justices that additional withdrawals would seriously injure their interests. Negotiating might make more sense than litigating under such conditions. In addition, by the early 1950s, basin states had spent years working together to better manage and protect the Delaware. In 1936, New York, New Jersey, Pennsylvania, and Delaware formed the Interstate Commission on the Delaware River Basin (INCODEL) to address water pollution and other cross-jurisdictional concerns. Inspired by this new spirit of collaboration among the basin states, in 1950 INCODEL officials unveiled an ambitious proposal to build a complex series of reservoirs for water supply and flood control along the Delaware. It was only when the INCODEL plan collapsed in late 1951 that the board opted to return to the Supreme Court to secure permission to dam the West Branch of the river to create the Cannonsville Reservoir.[40]

The new special master appointed by the Court to oversee the case, Kurt Pantzer, also strongly supported a negotiated settlement. "There were many good reasons," he noted, "to consider whether in a case involving the interests of so many communities and people law might now, without the loss of dignity, be the handmaid of science."[41] The three states (Delaware was not a party to the suit) submitted all testimony in writing, allowing for quick review by each party.[42] With little to show for its previous go-it-alone approach and a pressing need for more water from the Delaware, New Jersey entered into secret negotiations with New York.[43] After three months of meetings, New York agreed to a revised program of reservoir releases intended to ensure higher downstream flows, including the establishment of a flow measurement station in Montague, New Jersey. New York also agreed to permit New Jersey to divert 250 millions of gallons a day from the Delaware without building compensating reservoirs. In exchange for these concessions, New Jersey endorsed New York's increased diversion of Delaware River water.[44]

New Jersey's willingness to negotiate with New York fundamentally altered the Delaware discussions. Desperate for water to supply growing communities in the northern part of the state without compromising downstream flows, New Jersey struck a deal that it believed achieved both these objectives. Moreover, New Jersey would not have to incur the expense of building the compensating reservoirs that would otherwise have been legally required to divert water from the Delaware to other watersheds.[45] As William Schnader, a Philadelphia lawyer who represented Pennsylvania in the Delaware case, observed, New Jersey's defection forced Pennsylvania to negotiate rather than litigate: "When New Jersey gave its unqualified consent, Pennsylvania was more or less 'boxed in' because it cannot show any immediate damage for the reason that neither Philadelphia nor any other municipality in Pennsylvania has any present plans to go north of Trenton for its water supply."[46]

Pennsylvania used its remaining leverage to secure the best deal possible. Distrustful of New York, Pennsylvania insisted on the appointment of a river master to oversee the flows on the Delaware. Under what became known as the Montague Plan, New York City agreed to maintain a consistent flow at Montague, New Jersey, and to make additional summer releases based on the difference between the capacity of the watershed and the city's consumption levels. As New York's consumption increased, the volume of releases would decrease. The Keystone State also succeeded in reducing New Jersey's permitted diversion from the Delaware from 250 MGD to 100 MGD. Finally, Pennsylvania provided for its future water needs by securing the right to build a dam across the Delaware, and insisting that jurisdiction of the case remain with the Supreme Court.[47]

For its part, New York City cast itself in the role of beneficent nature improver. Far from putting the downstream states at greater risk of drought, it maintained, the revised release plan would deliver substantial benefits to New Jersey and Pennsylvania, including cleaner water, more reliable summer flows, and better recreation conditions. The board's chief engineer claimed that "The importance of the beneficial effects all along the Delaware river, resulting from New York's plan of development and the proposed diversion and releases, cannot be too strongly emphasized."[48] In response to Pennsylvania's concern that construction of the Cannonsville Reservoir on the West Branch would reduce flows in the main stem of the river, thereby permitting salt water to migrate upstream and contaminate Philadelphia's water supply (roughly half of which came from the Delaware), the board contended that the agreement it reached with New Jersey would have just the opposite effect. Because the Montague Plan required higher releases in the summer, when the river was at its lowest, building Cannonsville would send more water rushing downstream, keeping the salt front—the line at which the concentration of salt in the river rendered it unsuitable for human consumption—well downstream of Philadelphia's water intake.[49]

In June 1954, the Supreme Court approved the agreement proposed by the states. A region coping with the recent construction of the Rondout and Neversink Reservoirs and the imminent flooding of a valley from the impending completion of the massive Pepacton Reservoir would soon see another dam rise from a riverbed. After years of studies, panels, exaggerations, and negotiations, Cannonsville was on its way to becoming reality.

A Cold Reception

In many respects, the development of the Delaware system was a case of déjà vu. Waterworks construction unraveled the social and economic fabric that had held together both individual farms and entire communities, just as it had at Ashokan. New York City resumed work on the Neversink and Rondout Reservoirs in 1946, and initiated construction of the Pepacton Reservoir in 1947. George McMurray, a dairy farmer who lived several miles east of the Pepacton Reservoir, was one of the thousands of Catskill residents whose lives were turned upside down by the construction. Although New York allowed McMurray to retain title to his home and barn, in 1954 it seized his pastureland to prevent his herd from polluting a brook that flowed into the reservoir. The city initially rented McMurray's land back to him, but terminated his lease in 1961. His attorney, Herman Gottfried, neatly encompassed the farmer's case in a single question he posed to McMurray during his hearing before a damage commission: "So that when the City took your tillable land it practically took your farm?" Unable to nourish his cows, Murray retired from farming after thirty-six years.[50]

And, much as it had in the Ashokan region, reservoir and dam construction rearranged social and economic geography. Charles Webb ceased dairy operations on his farm when New York shuttered the creamery in the small town of Rock Royal, which would soon lie at the bottom of the Cannonsville Reservoir. For Webb, the construction of the reservoir amounted to a litany of inconvenience. His life used to revolve around the services available in Rock Royal, where he attended church, shod his horses, bought his groceries, and collected his mail. With the towns of Rock Royal and Cannonsville submerged, Webb spent his days making the long drive to Walton to attend church and purchase animal feed. What had once been a short jaunt of three miles to purchase sawdust for his barn now consumed much of a morning.[51]

The extensive recreational and ecological consequences of building the Delaware system began a new chapter in the city's reshaping of the Catskills. Laced with clean and fast-running streams, the Catskills had long been one of the nation's prized fishing destinations.[52] The original plans for the Delaware system envisioned dams on several of the river's tributaries, some of which were renowned for their excellent trout fishing. The decision to dam the Delaware's West Branch spared these smaller streams, cheering the growing legion of Catskill anglers.[53] Their joy was short-lived. By the mid-1950s, fishermen and other Catskill residents began to focus on disturbing changes taking place on the East Branch of the Delaware downstream of the newly completed Pepacton Reservoir.

Long known as an angler's paradise, the East Branch attracted fishermen from across the country.[54] Stanley Munro, a local real estate broker who testified at hundreds of commission hearings, claimed he once counted license plates from twenty-three states on cars parked along the stream. From spring to early summer, when snowmelt and moderate air temperatures kept the stream cool, fishermen descended on the East Branch to catch the trout that favored the stream. With the arrival of warm summer weather, trout sought out cooler waters, yielding to bass, which thrived when stream temperatures increased.[55] Fishermen lined the banks well into fall, pumping money into the local economy, staying either at private camps they purchased for recreational purposes, or patronizing the boardinghouses that catered to urban anglers.[56]

All of this changed after New York dammed the East Branch to form the Pepacton Reservoir. Rivers may be among nature's most unpredictable creations, but prior to the damming of the East Branch the stream exhibited some fairly consistent patterns. Periodic freshets cleansed the river, providing a good flow for most of the year, and removed silt and algae that entered from various feeder streams. The presence of the dam radically altered this flow regime. Absent the freshets, the East Branch relied on releases of water from the reservoir to purify itself. The irregular nature of these releases permitted debris from tributaries and silt from the reservoir to build up, clogging pools in the river that had provided crucial fish habitat.[57]

The releases had other undesirable effects, including making summer feel a bit like winter on some days. Because it was drawn from the bottom layers of the reservoir, where little sunlight penetrated, release water was exceptionally cold, often averaging around forty degrees. When frigid release water mixed with the warm summer air, it produced a dank, unpleasant fog, prompting locals to fire up their furnaces in the middle of summer. Residents who refused to heat their houses in summer discovered that the fog "causes mildewing, dampens bed clothing, gets into clothes and even loosens the wallpaper."[58] The city's release regimen appeared to turn the local climate upside down.

The releases also transformed the recreational character of the stream. In summer months, New York City made large discharges from the Pepacton to meet the required flow levels dictated by the Supreme Court decree, dramatically lowering stream temperatures. Prior to its damming, the East Branch had been a popular swimming destination. But as a local surveyor and engineer observed, releases rendered the stream "uncomfortable to wade in for any length of time."[59] Even one of the city's own witnesses acknowledged that he would inform a prospective homebuyer that it "is going to be too cold for you or your kids to swim in that river."[60] With the reservoir itself closed to swimming, local residents and summer visitors pondered a strange paradox: surrounded by water, they had nowhere to swim. Providing water for downstream states meant decisively altering the character of streams in the Catskills.

The intrepid few who braved the chilly waters shared the East Branch with a new cast of aquatic characters. One of America's premier warm water fisheries had, virtually overnight, become a cold-water trout stream. Much of the evidence concerning the dramatic changes on the East Branch emerged in business and property damage hearings held in connection with construction of the Pepacton Reservoir. Local residents and their witnesses contended that construction of the reservoir and the associated releases lowered their property values by degrading the recreational quality of the stream. Witness after witness, from aquatic biologists to those reared on the river, agreed that cold-water releases had destroyed the East Branch's smallmouth bass fishery. The releases simply made the river too cold for bass and other warm-water species to survive.

Edward Raney, a Cornell professor of fishery biology who testified on behalf of several local property owners, explained the ripple effect of releases. Cooler temperatures interfered with the reproduction patterns of warm-water fish. Forced to migrate in search of warmer temperatures, bass frequently deserted their nests, leaving their eggs vulnerable to deadly fungi. Due to the need to meet flow requirements at Montague, New York City frequently released enormous volumes of water in a short time span, causing water temperatures to plummet. In an eight-day stretch in June 1957, water temperatures in the East Branch fell from a comfortable seventy-two to forty-eight chilly degrees. According to Raney, the release program "was absolutely disastrous for the warm-water fishery in the East Branch."[61]

Confronted with such damning evidence, the city tried to turn the tables on its accusers, arguing that the release program created a vibrant trout fishery on the East Branch. Cooler temperatures certainly created a more conducive environment for cold-water species such as trout, but residents disputed New York's claim that it had simply replaced one fishery with another of equal value. Harry Darbee, who crafted custom flies for Catskill fishermen, insisted that the increase in trout populations did not translate into better fishing, because fish refused to take bait when temperatures plunged.[62]

Even when trout were biting, the benefit to the local economy could not compare with the prosperous era of the bass fishery. According to one witness, changes on the East Branch attracted bait fishermen, who eschewed Darbee's expensive artificial flies. As Stanley Munro explained, "They are not the monied type of fishermen. They do not have the funds, for example, to sit down in a bar room and drink scotch and soda."[63]

At its core, the dispute about recreation on the East Branch revolved around the irregularity and unpredictability of the release regimen. Although the city, under pressure from the state, made so-called conservation releases to improve stream conditions during the summer, the effect of these small discharges paled in comparison to the much larger releases it made to comply with downstream flow requirements. When natural flows were high, the city husbanded its supplies, making only the token releases required by the state. A state report on downstream releases on the East Branch in 1959 concluded that these irregular release patterns "relegated the river to a perpetual drought condition except during periods of springtime reservoir spill and during late summer (July–September) when the River Master needed large releases from Pepacton to maintain the . . . minimum flow in the main river at Montague."[64]

The combination of human interference, legal prescriptions, and variations in rainfall led to unpredictable conditions over the medium and long term, frustrating swimmers and fishermen. Looking back on the period from the early 1950s, when workers were busy erecting the dam that created the Pepacton Reservoir, to the mid-1960s, state officials discerned three distinct ecological phases. When the reservoir first began to store water, flows on the East Branch were quite low. Once the reservoir was filled in 1956, frequent releases boosted stream flow, which promoted development of a trout fishery, however tenuous and intermittent. With the onset of drought in 1960, conditions in the East Branch changed once again, creating a worst-of-both-worlds situation, according to state biologists: "Although large releases still occurred after 1960, the interspersion of high flows with low flow periods provided neither coldwater nor warmwater habitat in the East Branch."[65] Contrary to the confident predictions of BWS officials that city management would increase low flows during the dry summer months, improving on nature's fickleness, evidence from the first decade of the new release regimen suggested that humans lacked the capacity (or, perhaps more accurately, the desire) to balance large diversions of water for urban dwellers with the recreational and ecological needs of rural residents.

The demands of suburbanites were more straightforward: they wanted an ever-increasing supply of New York City's water. The city's legal obligation to permit communities throughout Westchester County to draw upon its reservoirs and aqueducts catalyzed growth and reshaped the suburban built environment. Just as it had inside the city and throughout the Catskills, the water system played a critical role in transforming the economy and landscape of New York's suburbs.

Urban Water, Suburban Growth

The state legislators from Westchester and Putnam Counties who negotiated the landmark bill permitting New York City to tap Catskill streams exhibited unusual foresight. Their constituents generally lived in small towns and villages and identified more with hardscrabble Catskill farmers than with the burgeoning metropolis. Most residents enjoyed ample water supplies drawn from local sources. The men they elected to represent them in Albany envisioned a different future for the region. Abandoning their traditional alliance with rural interests, in 1905 they struck a deal with New York: the city could appropriate water from the Catskills if it agreed to share the bounty with Westchester and Putnam. If the significance of the quid pro quo was somewhat murky in 1905, it had become clear by the early 1970s, when New York City supplied almost three-quarters of the water consumed by Westchester residents.[66] Reflecting on the legislation almost seventy years after its passage, a state commission neatly summarized the upshot of the 1905 pact: "The City became a regional supplier through a trade-off: the right to develop water in upstate counties, in exchange for the right of these counties to tap into its lines at cost."[67]

The first instance of Westchester County's reliance on New York's water network occurred in 1910, before any towns established permanent connections to the city system. A severe drought struck the state, prompting widespread water-use restrictions. With Catskill dams and reservoirs still in the early stages of construction, New York drew mostly on its Croton system. Its collection of large reservoirs provided a measure of safety unavailable to Westchester communities reliant on small reservoirs and groundwater supplies. In late September 1910, the city of Mount Vernon was on the verge of running out of water. New York agreed to permit Mount Vernon, which borders the Bronx, to tap into its supply. It also provided water to other Westchester County communities. A review of the county's water policy published in the 1930s gratefully acknowledged this critical aid: "As a matter of neighborly consideration, and without any legal obligations, New York City permitted the various water districts to draw upon the Croton supply and the emergency was relieved."[68]

The warm relationship between the city and Westchester did not last long. In the spring of 1911, New York City proposed repealing state laws restricting it from developing water sources in Putnam and Dutchess Counties. Fearful that another severe drought would occur before it could deliver water from the Catskills, municipal officials hoped to tap into a new source that could be developed relatively quickly. Its generosity toward Westchester in the drought of 1910 was likely an effort to curry the favor of legislators whose support New York needed to gain access to these water sources. The city's bid to overturn the state restrictions failed, leading it to adopt a new strategy. Instead of sharing supplies with Westchester communities, it did everything in its power to restrict their access to New York water.[69]

This stinginess reflected basic mathematics; with a population swollen by massive immigration, increasing per capita consumption, and a lag of several years before the arrival of Catskill water, the city feared that any diversion to Westchester would compromise its ability to deliver water to its own citizens. Its state legislators persuaded their colleagues to require the approval of the state conservation commission before allowing other communities to connect to its water system. In 1912, fed up with the inconsistent service provided by the community's private water company, Mount Vernon's citizens endorsed a plan to connect to New York's pipelines. The state rejected the request, arguing that Mount Vernon should further develop local water sources before tapping into New York's supply. Mount Vernon was irate. By refusing to supply water to Westchester, the Mount Vernon city historian observed, "New York City was seeking to welch on the promise it had made to secure the Catskill water." Mount Vernon led a vigorous lobbying effort in Albany to dismantle the legal obstacles that stood between Westchester and New York's water. Donning buttons reading "Westchester Wants Water," residents of Mount Vernon and other county communities prevailed—in 1916, the governor signed legislation permitting cities and towns in Westchester and other communities where New York had water facilities to connect to the water system without city or state permission.[70]

Access to New York's water helped spur the county's breakneck growth. Although most Americans think of mass migration to the suburbs as a post–World World II phenomenon, waves of suburbanization began to reconfigure Westchester shortly after the turn of the century.[71] By 1930, the county was home to over half a million people, almost triple the population in 1900. In dry years, New York City supplied almost half the county's water.[72] Some communities relied exclusively on city supplies, while others combined local sources with New York water. Access to New York's reservoirs and aqueducts enabled growth that might otherwise have been curtailed by the population's skyrocketing demand for water. As one water expert observed in 1934, "There is no real danger of shortage, thanks to the physical connections between the water districts and the New York City systems."[73]

In strategically situated communities in southern Westchester such as Mount Vernon, the availability of New York water encouraged denser and more vertical forms of development. In 1925, Mount Vernon established a connection to the city's Hillview Reservoir, located in the neighboring city of Yonkers. Because most of Mount Vernon lies more than a hundred feet below the level of Hillview, it enjoyed exceptional water pressure. The availability of a reliable source of water that required little or no pumping to ascend several stories above city streets spawned an apartment building boom in the 1920s.[74] Mount Vernon's economy and physical form began to resemble its neighbor to the south—the Bronx— much more than the rural towns of northern Westchester. In only two decades, its population increased by more than 60 percent. In 1947, longtime local newspaper reporter Philip Anderson reflected on the role that New York water had played in transforming the community: "The coming of Catskill water . . . became a strong talking point in the building up of Mount Vernon."[75]

Westchester's escalating demand for New York City water in the 1950s and 1960s reflected changing patterns of growth and the limitations of natural systems. At the most basic level, Westchester's increasing reliance on New York water was a result of a 30 percent increase in the county's population in the 1950s. The author of a 1955 report on the county's water system linked Westchester's population boom to the decreasing attractiveness of urban life: "Living conditions within the New York City area have—for nearly three decades—been so congested as to give rise to a gradual, at first, and now, an accelerated exodus of people seeking more pleasant living conditions."[76] In theory, a family who moved from the Bronx to a Westchester community with connections to the Catskill aqueduct did not increase the overall demand on the water system, because it consumed city water in both places. But the watering of shrubs, gardens, and lawns in the suburbs generally led to higher per capita water use.[77]

The changing nature of the county's business base also contributed to increasing water consumption. Major corporations joined the suburban exodus, leaving New York City for Westchester. Nestlé and General Foods led the way, in 1952 and 1954, respectively, followed by IBM, which moved its world headquarters to the Westchester hamlet of Armonk. Texaco, Union Carbide, and other firms soon followed. Although most of their Westchester operations involved administrative and service tasks, not industrial activities that consumed large volumes of water, the sheer scale of these corporations significantly boosted water consumption in many communities.[78]

The combination of increased residential and commercial development threatened to compromise many of the water sources that communities had relied on for decades. A committee formed in 1951 to study Westchester's water system highlighted the tension between growth and environmental protection: "As the population in Westchester continues to grow; as local reservoirs will increasingly lack sufficient storage capacity; as the reservoirs may have to be abandoned because of watershed contamination; just so, it seems inevitable the county will become more dependent upon New York City water, or will have to provide a complete new water supply system of its own."[79] This analysis was prescient. By appropriating the county's two best sources—the Croton River and its tributaries and the Bronx River—New York effectively doomed the prospects for a Westchester water system. Westchester officials first considered building a county water system in the 1920s, but the high cost of tapping distant supplies deterred them.[80] By the early 1960s, reconciling growth and watershed protection proved increasingly difficult.

The vulnerability of the Croton system to pollution exemplified the challenges of balancing watershed protection and economic development. Depending on their location, Westchester cities and towns tapped into different parts of New York's water network. Communities along the Catskill and Delaware Aqueducts enjoyed access to relatively pristine supplies delivered under high pressure. Some cities, such as New Rochelle, had multiple connections to New York's water network. When consumers complained about poor water quality, New Rochelle's water company ceased drawing water from Croton aqueducts, and the complaints stopped.[81] Towns that pumped water directly from one of the Croton reservoirs into their distribution systems had to construct filtration plants to ensure water quality.[82] In the 1970s, many Westchester towns followed New Rochelle's lead, severing their connections to the Croton system in favor of Catskill water. A combination of inferior water quality and high energy costs (Croton water generally had to be pumped before delivery, but Catskill water did not) led them to abandon the water that New York City had appropriated from them more than a century earlier.[83]

Ironically, around the same time that Westchester towns and villages turned away from Croton water, they began to enjoy a new recreational asset courtesy of the Croton system. The availability of additional water from the Catskills enabled New York City to suspend use of the Old Croton Aqueduct in 1955. While the aqueduct was in service, the city maintained a right of way to protect it from damage. With no plans to resurrect it for water delivery, the twenty-six-mile route of the aqueduct in Westchester could be put to another use. In 1968, the state acquired the Westchester section of the aqueduct for use as a park. The park—essentially a wide trail—stretches from the Croton Dam in northern Westchester to the border with the Bronx, and is extraordinarily popular with runners and walkers, who enjoy its historic homes and scenic views.[84]

The conversion of the Old Croton Aqueduct right-of-way symbolized the protean quality of the Croton system in rapidly suburbanizing Westchester. By the late 1960s, most Westchester residents viewed the Croton system not as a human construction that displaced farmers and small-town residents or as a significant water source for their growing communities, but as an indispensable recreational asset that provided refuge from the stresses of urban and suburban life. The string of reservoirs provided convenient access to excellent recreational fishing that complemented the walking, hiking, and biking of the Old Croton Aqueduct trail. The reservoirs and surrounding land buffer also sheltered northern Westchester from the more concentrated development that came to characterize the southern portions of the county. The Kensico Reservoir and the city-owned woods that surrounded it connected with other large parcels such as Grasslands Reservations, site of many county facilities, forming a curtain of sorts between southern and northern Westchester.[85] With the opening of the Bronx River Parkway in 1925, whose original 15.5-mile route linked the northern Bronx and Kensico, picnicking on the reservoir's beautifully landscaped grounds became a popular Westchester tradition. In both urban centers and rural villages, New York's water system played a critical role in shaping the county's natural and built environments.

Thanks to the vision of state legislators, Westchester hitched a ride with New York City on the Catskill water gravy train. Beginning in the nineteenth century, the county had been forced to surrender its best water sources to the Goliath on its doorstep, but by cutting a deal that allowed New York City to tap Catskill water, Westchester charted a course that would reshape its economic and physical landscape. Westchester gained access to high-quality water with fairly minimal capital expenses. A plentiful supply of water promoted economic development, transforming the county into a bustling center of commerce during the week. On weekends, its residents could retreat to one of the many Croton reservoirs to fish, head to Kensico to cool down in the spray of the massive aerators, or take a vigorous walk along the trail of the Old Croton Aqueduct. As long as the weather cooperated, Westchester enjoyed abundant supplies of high-quality mountain water.

The Drama of Drought

Unprecedented in duration, severity, and scope, the drought of 1965 dramatically raised the stakes of the battle for water in the Mid-Atlantic region. As U.S. Secretary of the Interior Stewart Udall observed, no one could remember anything like it: "Neither weather nor streamflow records of the past century and a half, nor older historical writings, newspapers, nor other records, show its equal."[86] The region had been in the grip of a drought since 1960, but a rainy spring in 1964 raised reservoir levels, prompting officials to suspend most mandatory water conservation measures.[87] Unfortunately, the relief proved short-lived. From November 1964 to May 1965, precipitation in city watersheds was only half of the historical average. Equally distressing, much of the rain that did fall on the Catskills was absorbed in the root systems of parched trees and shrubs before it could make its way to the reservoirs. A remarkable 80 percent of precipitation usually drained into the reservoirs; by early 1965 only 56 percent did.[88] The combination of reduced precipitation and diminished runoff left New York's reservoirs perilously low in the spring of 1965.

Relative to the Mississippi and many Western river systems, the Delaware basin is of modest size; its significance stems from its geographical location—in the heart of America's most densely populated region. In the mid-1960s, 10 percent of the nation's people resided in the Delaware's drainage basin. In addition to providing one-third of New York City's water and half of Philadelphia's, the river and its tributaries supplied many communities in New Jersey.[89] Despite the heavy demands on the basin's resources, most experts believed that the river could meet the needs of the region.

The drought crisis of 1965 represented the first major crisis for the Delaware River Basin Commission (DRBC), the organization the four basin states had established in 1961 to improve conditions on the river.[90] The governing structure of the DRBC was unique: its five official members consisted of each basin governor and a federal representative. This federal-state body was the first of its kind.[91] Although New York City did not have its own seat on the DRBC (New York State represented its interests), it secured a crucial concession from the other basin states in exchange for agreeing to creation of the commission: they reaffirmed the city's right to divert the Delaware. Wary of the increasingly interventionist tendencies of the federal government, the states viewed the DRBC as the most promising vehicle for cooperative development of the basin.[92]

Although it possessed much more clout than its predecessor INCODEL, the DRBC lacked the authority to mandate metering or other water conservation measures in New York City. The city's water conservation program continued to lag well behind other major cities. Despite the first detectable long-term decrease in the city's population in centuries, consumption escalated because of leaky water mains, faulty plumbing fixtures, and unmetered buildings.[93] Even with the infusion of water from new reservoirs completed in the 1950s, the prolonged drought posed a grave threat to New York's water security. As one dry day followed another in the spring of 1965, Mayor Wagner had little choice but to ban the sprinkling of lawns and washing of sidewalks. A few weeks later, the city shut off most public fountains, and required restaurants to provide water only upon request.[94] In July, the Department of Water Supply took to the skies. Two blimps hovered over New York flashing the water conservation gospel: "Take Brief Showers Instead of Baths," "Stop Running Hot Water on Dishes."[95]

The city's actions to expand its supplies were much more controversial than emergency conservation measures. In the middle of June, with no advance warning to the other basin states, it ceased releasing water from its Delaware reservoirs back into the river, a flagrant violation of the 1954 Supreme Court decree requiring New York to release water to sustain stream flow during the dry summer months. Suspending its releases gave New York more water for its own citizens and slowed the depletion of its reservoirs. It also outraged New Jersey and Pennsylvania because it reduced the volume of water flowing downstream.

Philadelphia was particularly alarmed because any reduction in the volume of fresh water flowing down the Delaware allowed salty ocean water to migrate further upstream, putting its water supply in jeopardy.[96] By early May of 1965, the salt front had advanced fifteen miles farther upstream than the previous year.[97] Growing increasingly nervous with each dry day, officials in Pennsylvania worried that New York's City's refusal to honor its release requirements put Philadelphia in grave danger. The nation's largest city had defied the Supreme Court, and Philadelphia's two million citizens appeared to be paying the price. Maurice Goddard, Pennsylvania's top environmental official and DRBC representative, charged New York City with "pirating" water.[98] The suspension of releases directly contradicted assurances made by the Board of Water Supply during the Delaware hearings of the 1950s that construction of the Cannonsville Reservoir would protect Philadelphia's water supply.

The conflict between New York and Philadelphia pitted cities with starkly different water supply philosophies against each other. Large rivers ran through both cities: the Hudson marked Manhattan's western edge; the Schuylkill and Delaware threaded Philadelphia. Philadelphia had at various times over the last half century contemplated tapping mountain streams for its water supply, but, wary of the great expense, continued to rely on its two industrial rivers.[99] A modern filtration system and extensive chemical treatment ensured its citizens a safe water supply. Philadelphians had long wondered why New York did not tap the Hudson. They scoffed at New Yorkers' insistence that the river was too polluted. The Hudson at Hyde Park, observed a Philadelphia journalist in 1952, "is almost champagne by contrast with the Delaware water which Philadelphia drafts through the Torresdale intake."[100]

Philadelphians also argued that New York's water crisis stemmed in large part from its reluctance to meter all water users. Philadelphia began requiring meters on all new water connections in 1917, and in the 1950s placed meters on all of its older buildings. Universal metering was a resounding success, reducing water consumption by almost 20 percent.[101] The fact that "water spend-thrift New York," which adamantly refused to meter residential properties, was putting its more prudent neighbor at risk galled Philadelphians. Invoking Aesop, one Philadelphia newspaper editorialized that New York was "a little like the grasshopper who, having fiddled all summer, starts robbing the industrious ant during the long winter."[102]

With its members bickering on the front pages of the daily papers, the DRBC moved to resolve the crisis. It declared a drought emergency in early July. With the consent of both the city and New York State, the commission temporarily reduced both the amount of water New York City could divert from the Delaware's tributaries and the required volume of its downstream releases. By flexing its legal muscles—the commission was the only institution authorized to modify the Supreme Court decree—the DRBC defused the short-term political crisis.[103]

But by the dog days of summer, with substantial rain a distant memory, New York began to get desperate. Some observers suggested resurrecting the Hudson River pumping plant that the city had constructed in the Dutchess County hamlet of Chelsea during the drought of the early 1950s. After the rains returned, New York City had dismantled the plant and sold the land. As late as May 1965, the water commissioner continued to deride the Hudson, insisting that the city would not tap such a polluted source even in the midst of the severe drought.[104] Two months later, Mayor Wagner overruled him, declaring that New York would rebuild the plant. Perhaps pleased that his much-derided Hudson was now perceived as a crucial asset, Abel Wolman tartly remarked, "No Californian would have been foolish enough to abandon Chelsea."[105] The mayor even floated the possibility that New York would resort to rainmaking, and once again Wallace Howell offered his services.[106] Fifteen years and four reservoirs later, the nation's largest city again considered betting on cloud tinkering to survive a drought.

The most promising short-term approach for increasing supplies—suspending its downstream releases altogether—would require New York to cooperate with its neighbors. By early August, the city began to issue dire warnings, declaring that it would run out of water by February.[107] On August 11, President Johnson summoned the region's governors and mayors to Washington to discuss the drought. He then dispatched water crisis teams led by Interior Secretary Udall to the major cities in the drought region, and announced plans for a water resource study of the region.

Despite the president's insistence that water supply was primarily a local responsibility, the federal government forcefully intervened to resolve the crisis. After touring the region's cities, Udall convened a meeting of the DRBC in Washington and proposed a "water bank" as the solution to the mounting crisis. New York would be permitted to store (escrow) in its reservoirs the water it would normally release downstream. The DRBC would control this banked water and determine at a later date if it should be used for releases or diversion to New York. To compensate for the suspension of releases from city reservoirs, the agreement called for increased releases from private power reservoirs on Delaware tributaries and from an Army Corps of Engineers flood control reservoir. In addition, New York State agreed to transfer water stored in a lake on its border with New Jersey into reservoirs for the city of Newark.[108]

Udall's proposal held clear benefits for all involved. The suspension of releases provided New York a much-needed cushion, allowing it to conserve the equivalent of an entire day's water use every week. Meanwhile, New Jersey could tap federal funds to pay for the transfer of water from New York State into Newark's reservoirs. The plan also allayed Philadelphia's concern about the encroaching salt front by increasing overall flows on the Delaware and guaranteeing that the federal government would relocate Philadelphia's water intake further upstream if necessary. Implementation of the water bank plan effectively ended the crisis. Although water levels in city reservoirs did not return to normal until the end of 1966, by the fall of 1965 the salt front was moving downriver, and New Yorkers no longer feared a water shortage.[109]

The drought resurrected long-standing concerns about New York's outmoded water management practices. The high stakes of the drought and the upcoming mayoral election (Robert Wagner was stepping down after more than a decade in the post) turned the water system into political fodder. Democratic mayoral candidate William Fitts Ryan made poor water management a major theme of his campaign; after uncovering a gaping leak in the Central Park Reservoir, he demanded the resignation of the water commissioner.[110] Interior Secretary Udall, the president's point man on the drought, ridiculed the city's water practices, contrasting Philadelphia's modern, unified water department and forward-thinking approach with New York's outdated pricing and conservation strategies. Even President Lyndon Johnson chimed in, observing brusquely, "There is not much I can do about the third of Bob Wagner's water that we don't know where it is going."[111]

Put on the defensive, Mayor Wagner justified the city's reluctance to meter residential consumption as the product of its advanced approach to social welfare.

Taking a cue from previous mayors, Wagner cast New York's policies as uniquely progressive: "We have been very proud in this city of the fact that water has been almost as free as air. By this means we have encouraged the maximum beneficial use of it."[112] Firmly rooted in a pre-ecological era, Wagner refused to acknowledge that profligate water use by New Yorkers adversely affected others in the river basin, whether in Philadelphia or the Catskills, and blamed the weather for the city's plight. Others were not so sure. Following Secretary Udall's visit to New York, during which he scolded the city for its leaky infrastructure and lack of metering, the *Times* ruefully concluded that "if we now sit on the edge of disaster, a four-year drought is not wholly responsible."[113]

At its core, the water conservation debate boiled down to one critical question: Should New York install meters and charge its residents based on the volume of water they consumed? In most American cities, metering water use generated little controversy. But in New York, metering remained politically explosive. The city's unique residential structure meant that landlords—the constituency paying the majority of water bills—exercised disproportionate weight in discussions about water supply. Even in the midst of an unprecedented drought, property owners continued to unanimously oppose the installation of water meters. John Berger of the Greater New York Taxpayers Association explained the problems posed by the typical apartment dweller: "He has always considered water as a God-given gift which comes to him from heaven for his free and unlimited use; and it is the sheerest kind of folly to believe that the mere installation of a registering device in the basement of apartment or tenement houses in this city will exercise some form of magical restraint upon the millions of water users who have always been profligate with water."[114]

Landlords feared that the tenuous connection between water rates and consumption would result in exorbitant bills under universal metering. Rates could only serve as a conservation tool if they rose with increases in consumption. However, pipe arrangements in most apartment buildings made installing meters in individual apartments impractical. As Berger noted, conservation proponents envisioned placing a meter in the basement of each building. Although individual tenants would not have a direct incentive to reduce their water use, conservation advocates insisted that metering would curb overall water consumption. By measuring overall water usage in a building and charging according to consumption, a common meter system would likely induce landlords to repair leaking toilets and sinks.[115] Eager to preserve low rates, landlords dismissed the problem of leaky fixtures, arguing that it represented only a sliver of overall water use.

These issues came to the fore when Mayor Wagner appointed a group of municipal officials to study the water system and make recommendations. The officials echoed the recommendation of previous panels in calling on New York to implement universal metering. The panel initially proposed coupling universal metering with rate increases of 25 to 100 percent, reasoning that New Yorkers would only curb their water use when they had a strong economic incentive to do so. Property owners savaged the proposal in the press, prompting the panel to drop its insistence on sharply increased rates in its final report. Nevertheless, the report's forthrightness distinguished it from previous analyses of the water supply. Its authors even highlighted the ecological and political connections between New York and its watershed neighbors: "We go far to seek our water supply. We share use of water resources with other communities of our region, the majority of whom are metered, and we must cooperate with them as we expect them to cooperate with us."[116]

Seemingly backed into a corner—New York was still officially under a drought emergency and the report had been prepared by several high-level municipal officials—Mayor Wagner found support from an unlikely source: Abel Wolman. The mayor created yet another expert panel to evaluate the water system. Wolman and his fellow panelists agreed that tracking consumption in different types of buildings would improve management of the water system, but contended that universal metering would cost more than it would save. The panel also recommended immediate house-to-house inspections and changing the rate structure to discourage extravagant water consumption.[117]

Mayor Wagner selectively embraced the Wolman panel's suggestions. He issued an executive order requiring most municipal agencies to install meters. The Department of Water Supply strongly encouraged hospitals, museums, and other institutions that were by law exempt from water charges to install meters to ensure they were not abusing their free water supply. But the mayor shunned all measures that had the potential to alter the essential philosophy or incentive structure of the water system. Most significantly, he decisively rejected universal metering. Consumption would still have no bearing on water charges, which, the mayor emphasized, would be set as low as possible.[118]

The worst drought of the century was not severe enough to overcome entrenched opposition to universal metering and water conservation. The ideology and political arrangements that had long thwarted any change in the city's water policy remained intact. In 1960, political scientists studying New York's water system concluded that low water rates had become "almost a constant of nature." Amazingly, a five-year drought did little to change this mind-set.[119] By 1967, the return of the rains had washed away hopes of substantive reform. John Lindsay, who succeeded Wagner as mayor, proposed universal metering, but his bill languished in the city council. Widely recognized as a city in crisis by the middle of the 1960s, New York and its elected officials focused on more pressing problems of poverty, racial strife, and moral decay.[120] Rain filled up the Cannonsville Reservoir, providing New York with an additional margin against drought. The worst fears of a prominent Philadelphia good-government organization, expressed in September 1965, had been realized: "A danger is that, once the present drought is over, public officials and concerned citizens will quickly forget the entire experience and return to an attitude of normalcy until, once again, a crisis occurs."[121]

The End of an Era

The year 1967 represented an important milestone in the history of New York City's water supply: after forty years of planning, delays, and construction, the city's Delaware water system was finally completed. After this date, the city did not seek additional permanent water sources. New York did spend billions of dollars on a variety of capital projects in the final decades of the century, including the construction of City Water Tunnel No. 3 (scheduled for completion in 2020, the tunnel will enable the DEP to take one of the two existing distribution tunnels offline to make long-overdue repairs), refurbishing of dams, and replacement of much of the city's antiquated private plumbing. Nonetheless, 1967 marks a sharp divide between expansion and system maintenance, between the end of an abiding vision and the slow birth of a new approach to water supply management.[122]

Some of the beneficiaries of these waterworks projects did not reside within the five boroughs. With each passing year, an increasing percentage of Westchester County residents came to rely on New York City's network of reservoirs and aqueducts to water their lawns and quench their thirsts. In the nineteenth and early twentieth century, the construction of New York's Croton system destroyed communities and dramatically remade the county's northern fringe. In subsequent decades, access to city water enabled growth throughout Westchester, but especially in its southern reaches, where rapid urban and suburban development transformed the landscape. Westchester's residential and commercial boom rested on a solid legal and ecological foundation: access to New York City's abundant supply of Catskill water.

Of course, abundance was not an absolute asset; what mattered was the balance of supply and demand. New York City's vulnerability to droughts stemmed from a variety of causes, including decades of population growth and delays in constructing the Delaware system. However, even after the population stabilized in the mid-1960s and the city completed the Delaware system, droughts remained a threat. A pronounced hostility to water conservation was responsible. By any reasonable standard, consumption was exorbitant. Failure to meter residential properties, low water rates, and poor maintenance of aqueducts and mains led to substantial waste. The ability to draw on the Catskills for additional supplies delayed the day of reckoning.

Nevertheless, the city's decision to substantially expand its upland supply system even as its population stagnated puzzled many outside observers. Why did New York opt to spend hundreds of millions of dollars on new supplies rather than tens of millions on conservation? Why did the inertia of received wisdom prove so powerful even in the face of substantial opposition? The answers to these questions illuminate both the politics of the immediate postwar decades and the forces and attitudes that shaped the city's water policy until the 1990s.

City government may have lacked the consistent support of the fourth estate, but it enjoyed the backing of a more significant, if less vocal, ally: the law. At both the state and federal levels, the criteria used to evaluate the legality of the Cannonsville project were narrowly defined. In this sense, the law performed its requisite role in democratic society of boiling down a multilayered, complex conflict into a more simplified reckoning of competing goals. In its decision approving New York's application to build the Cannonsville Reservoir, the state noted the objection of local residents who claimed that "mere financial remuneration can never properly compensate owners for such losses as disruption of old friendships and neighborhood ties." Although they conceded the point, the state water commissioners went on to observe that such sentimental considerations played no part in their decision. Their charge was simply to determine whether the project was justified by public necessity.[123]

Cannonsville was a classic example of what political scientist James Scott calls bracketing—defining the objective of public policy so narrowly that the benefits of alternative approaches are not factored into the decision-making process.[124] Drawing on the Hudson would have preserved the social fabric and ecology of the Cannonsville region, encouraged water conservation, and reduced the negative effects of dams on downstream states. The strict utilitarianism of officials at all levels of government reflected the lingering institutional power of Progressive Era administrative entities such as the State Water Power and Control Commission and the Board of Water Supply. Pitted against New York City's multitudes, the people of the Catskill Mountains had no realistic hope of stopping the Cannonsville project. In an era before environmental impact statements and widespread recognition of the negative repercussions of large infrastructure projects, Catskill residents lacked the legal leverage to effectively resist the project. Bereft of allies in state government, watershed residents could not gain political traction. Just as impoverished city dwellers lost their homes in the name of the greater good of urban renewal, so were Catskill residents displaced for the benefit of millions of New Yorkers.[125]

The involvement of the federal government did not substantially alter prevailing legal and ecological framework assumptions. The Supreme Court merely affirmed the deal struck by Delaware basin states, a deal predicated almost entirely on securing equitable access to the Delaware for water supply purposes. Ecological concerns were notable only for their absence. With New York State firmly behind the proposal to build a new reservoir, the interests of Catskill anglers and residents went entirely unrepresented.

Financial factors also reinforced the water supply status quo. The ease of issuing bonds to cover the cost of waterworks construction obscured the enormous cost of building new reservoirs and tunnels. The city's ability to pass on the cost of developing the Delaware water system to succeeding generations reduced the attractiveness of less expensive supply options such as the Hudson. A pumped system like the Hudson would also increase operating costs, which were funded out of the annual operating budget. Moreover, bonds for water supply projects were exempt from a state law limiting a city's indebtedness to 10 percent of its real estate valuation. Issuing hundreds of millions of dollars in bonds for water projects therefore posed no immediate opportunity cost to the city. Bonds represented, in every respect, the path of minimal financial resistance.[126]

Crucial as these legal and financial considerations were, it was the outsize influence of the BWS and its vision of an abundant, inexpensive, and unmetered water supply that ensured that New York would continue its incursions into the Catskills. Cynics viewed the board's predilection for grand mountain dams and reservoirs as a ploy to maintain employment and prolong its very existence.[127] As long as the city continued to build massive new water projects, the three commissioners and the army of engineers and clerks who worked for them had no reason to fear for their jobs. These critics grossly underestimated the organization's sense of mission and uncritical confidence in its own vision.[128] Pumping dirty water from the Hudson held no appeal for engineers who prided themselves on masterminding the construction of complex dams and reservoirs. As one board critic observed, "The entire philosophy of the department is to build and build dams. They'd be terrified if they had to build a filter plant."[129]

Political, legal, and economic structures reinforced the existing paradigm of upland source expansion and inattention (bordering on willful neglect) to water conservation. The dams and reservoirs that were the products of this approach continued to shape the city, Westchester County, and the Catskills for years to come, but in the absence of a sustained government commitment to environmental protection, ecological considerations played little role in the debate over the city's water system in this period. This would soon change, however, with the rise of organized resistance to New York's watershed practices.

CHAPTER FIVE

The Water System and the Urban Crisis

On balmy summer evenings in the 1980s, New York City's reservoirs—past and present—were the place to be. Five decades after it had been filled to create Central Park's Great Lawn, the former reservoir site hosted the biggest concerts in New York's history. In 1980, more than three hundred thousand people came to hear Elton John; two years later, an even larger crowd packed the lawn for Simon and Garfunkel's historic reunion concert. The New York Philharmonic and the Metropolitan Opera regularly entertained tens of thousands on the lawn. With the Sheep Meadow closed for restoration, the Great Lawn, "one of the few great, dramatic open plains in New York," became the city's outdoor living room.[1]

Over a hundred miles northwest of Central Park, the Pepacton Reservoir also drew large crowds. With the fading of daylight, fishermen took to their rowboats (the only type of boat permitted on city reservoirs), their rods and lanterns in tow. Their lights attracted baitfish, which in turn lured the brown trout that flourished in the reservoir. A lucky fisherman might catch a fifteen- or twenty-pound brown trout in Pepacton's pure waters.[2] An angler described the tranquil setting: "In years past Huntey Hollow has looked like a little floating village, with Coleman lanterns from several dozen boats flooding the night."[3]

These pleasant scenes contrasted sharply with other, more worrisome developments. Fire hydrants, their caps removed by sweltering residents, flowed for days or even weeks into the streets. Critics, internal and external, began to question the quality of the water New York City delivered to its residents. As the focus shifted from supply expansion to network operation, the system's shortcomings became glaringly evident. The city had long relied on its ability to store pristine mountain water in large reservoirs for several months to ensure the quality of its supplies. By the 1980s, some officials recognized that nature needed a little help.

Socioeconomic changes complicated the transition to a more proactive water management strategy. Like cities from Los Angeles to Washington, D.C., New York suffered a profound social and economic downturn in the 1970s that threatened its ability to deliver a range of public services.[4] New York famously skirted the edge of bankruptcy in the mid-1970s.[5] The effects of economic contraction on the water system were felt both in New York—in the form of spilling hydrants, broken water mains, and skyrocketing water consumption—and in the system's watersheds, where the city fell behind on its property tax payments and allowed the bridges and roads that crossed and encircled its reservoirs to fall into disrepair.

Overwhelmed by the day-to-day challenges of delivering water to New York and other communities, officials largely ignored the implications of the changing Catskill landscape for water quality. In much of the region, the trend was residential growth, not decay. As the appeal of the city dimmed, second-home construction in the mountains boomed as old-time resorts gave way to new residences.[6] By the late 1980s, the executive director of a Catskill conservation organization had become alarmed: "Every resident of the greater New York Metropolitan Area, it seems, wants to own a piece of land in the country, and they are buying lots in the Catskills almost as fast as they are being subdivided."[7] Subdivisions were indeed sprouting in new places, but their growth obscured the long-term decline in tourism and agriculture, the pillars of the region's economy.

New York also began to confront a powerful new force: the environmental movement and the increasing prevalence of ecological thinking.[8] Beginning in the mid-1960s, the state and federal governments launched an unprecedented program of environmental conservation and regulation. Under Governor Nelson Rockefeller, New York emerged as a national leader in tackling the problems of pollution and economic growth.[9] The establishment of national standards for air and water quality, coupled with significant federal funding to help states meet these targets, transformed environmental regulation in the United States. Inspired in large part by citizen action, the new regime of environmental law provided citizens access to the courts to challenge government interpretation and enforcement of antipollution statutes. Charged with implementing most of the new regulations, many states—including New York—developed sophisticated environmental bureaucracies to protect natural resources and habitats.[10]

The underlying scientific and philosophical premise of the environmental movement was the belief in ecology—the notion that natural systems interacted in complex ways that could be disrupted by human interference. Ecological principles had the potential to radically alter humans' relationship with nature. Eager to demonstrate their environmental credentials but wary of undermining prosperity, politicians frequently sought technological solutions to problems whose roots lay in technology run amok. The cornerstone of the Clean Water Act was the establishment of a national permit discharge program that required mandatory adoption of the best available pollution control technology. The rise of ecological thinking did not topple technocracy; rather, technocracy subsumed environmental concerns under its mantle.[11]

This technological approach reflected the central role played by the middle class in the mainstreaming of environmental thought and action. Historians have identified increasing affluence as critical to the spread of American environmentalism. Ensured of plentiful food and comfortable shelter, middle-class Americans turned their attention to the quality of their lives and natural surroundings, and found them wanting. They demanded that government act to neutralize the unpleasant by-products of industrial capitalism: polluted air and water, the spread of toxic chemicals, and traffic jams.[12] As political scientist Richard Andrews observed, the intertwining of distinct strands of environmental concern among America's middle class created a potent new political force: "Public concerns about . . . ecological effects fused with those of organized nature-preservation groups and with more general public demand for outdoor recreation to form a mass movement of unprecedented strength and political diversity."[13]

Although watershed residents shared urban dwellers' enthusiasm for outdoor recreation, they generally viewed environmentalists with skepticism bordering on disdain. Confronted with declining agricultural and tourist sectors, they welcomed economic development and staunchly resisted environmental constraints. The urban roots of environmentalists led many Catskill residents to blame New Yorkers for the push to regulate development in the region. One irate dairy farmer sarcastically advocated invading the metropolis: "People from New York come to the Catskills to hunt, fish, ski and vacation and become immediate experts on how the Catskills should be managed. Let's reverse the situation and send a couple of Delaware County farmers to New York City. First they could plow up center field of Yankee Stadium and plant corn; after all, it belongs to all of us."[14] Local sportsmen, despite their resistance to outside interference, drew on cutting-edge scientific and ecological research in their efforts to reform what they viewed as New York City's disastrous downstream release policy.

In many ways, the degradation of the water system in the 1970s and 1980s reflected the more general sense of urban decay. Fed up with rampant crime and deteriorating municipal services, hundreds of thousands of residents left the city in search of a better life.[15] As its hordes of fleeing citizens indicated, New York City no longer controlled its own destiny; larger forces began to impinge on its ability to effectively maintain and operate its water system.[16] Catskill residents and the state and federal governments proved less willing to succumb to the city's demands. A post-imperial city, New York confronted a decidedly less favorable political and environmental climate.

The convergence of these internal and external pressures defined this critical transition period in the city's water history. New York struggled to adjust to a new reality where operational competence and consensus building began to loom larger than dam walls and massive reservoirs. Perhaps the clearest indication of the changed circumstances was the growing chorus of concern over the quality of New York's water supply. Long the pride of the system, water quality now threatened to become its Achilles heel.

The Water System Buckles

Nature's beneficence and astute planning by the Board of Water Supply had long preserved the quality of New York City's drinking water. The steep and heavily forested terrain of the Catskills inhibited large-scale development and provided abundant water runoff. New York stored these high-quality supplies in massive reservoirs, where a combination of exposure to sunlight and natural water movement over long storage periods—often up to several months—further purified them. Experts ranked the water as among the world's best.[17]

Although increasing development pressures in the Catskills threatened to degrade water quality, shoring up New York's dire financial condition far outweighed protecting its water sources. In 1977, the BWS even considered selling some of its reservoir buffer lands to replenish city coffers.[18] By the 1970s, New York City acknowledged it would need to construct a filtration plant to protect water quality in the Croton system; decades of suburban development left it no other option.[19] The population of Westchester County, home to much of the Croton system, had tripled since New York completed the system in 1911. As the cost of living in Westchester skyrocketed in the 1980s, development migrated north to Putnam County, the more rural portion of the Croton watershed. A watershed chosen in part for its remoteness from New York now posed a threat to water quality. By the 1980s, critics began to wonder whether the Catskill and Delaware systems would also require filtration due to increasing development in the watersheds west of the Hudson River.[20]

The newfound concern about water quality arose in a strikingly different institutional context. In 1978, New York City received authorization from the state legislature to eliminate the Board of Water Supply. With no immediate plans to tap new supplies, the board was deemed an anachronism that New York could ill afford. Its responsibilities, which consisted primarily of trying to kick-start construction of City Tunnel No. 3, were absorbed by a new agency, the Department of Environmental Protection (DEP). The uniformity of purpose and institutional autonomy that had defined the BWS disappeared overnight. The DEP, a traditional municipal agency riven by conflicting allegiances and a lack of resources, assumed control of all aspects of the water system.[21]

An incident in the summer of 1978, just months after the DEP assumed control over the water system, exposed the weaknesses of even the most basic departmental operations. Mayor Ed Koch visited the Brooklyn neighborhood of Bushwick, where a stream of water from an open fire hydrant gushed down a major street. Neighborhood residents had alerted the DEP weeks before, to no avail. More distressingly, DEP employees had inspected the sewers on the street a few days before Koch's visit, but ignored the open hydrant. In his letter to DEP Commissioner Francis McArdle, a clearly frustrated mayor sought reform through sarcasm: "I was intrigued by the thought that complaints of this nature to city agencies could go unattended for 30 days."[22]

In his response to the mayor, McArdle recounted the difficulties posed by a dysfunctional agency and recalcitrant citizenry. Rapid staff turnover and poor training, compounded by a lack of communication between different parts of the agency, lay behind such incidents.[23] Even when the DEP did act, McArdle argued, citizens often undermined their efforts. Emergency crews repaired over a thousand leaking water hydrants during a severe drought in 1980; within two days, overheated New Yorkers managed to reopen 20 percent of them.[24] By the mid-1980s, the DEP had improved its overall operations and reduced its average response time for leaks from eight weeks to three. Leaking and broken hydrants no longer symbolized the chaotic, decaying city.[25] As New York's financial position stabilized, the city sought to make up for lost time by investing heavily in water mains and trunks and restarting construction work on City Water Tunnel No. 3.[26]

Confined mostly to the city itself, these improvements had little bearing on the most serious problem confronting the water system—the degradation of its upstate sources. Although the DEP employed scores of people in the Catskills, it did virtually nothing to monitor the quality of water at the source. Peter Borrelli, who served as executive director of a prominent Catskill conservation organization in the 1970s, derisively recalled the city's narrow focus: "In those days, they were still plumbers. They were in the plumbing business." A lack of resources and emphasis on water delivery over watershed protection consigned New York City to the role of passive bystander. Even as a development boom engulfed the Catskills—fueled in part by urban dwellers who, in the energy crisis years of the 1970s, discovered that having a second home in the Catskills was cheaper than driving all the way to New England—the DEP failed to act.[27]

A small cadre of agency employees began to worry about the consequences of inaction. Vincent Coluccio was among them. He originally came to the DEP in 1980 as an intern, later becoming a full-time employee. A doctoral student in public health who took classes at night, he spent his days trying to apply what he learned to the water supply system. Coluccio focused on what he perceived to be the greatest water quality threat: the scores of municipal sewage treatment plants in city watersheds that discharged their effluent into streams that emptied into the reservoirs. He was shocked by what he discovered. New York City was doing "essentially nothing" to monitor or control sewage discharges at the sources of its unfiltered water supply.[28]

A shortage of money and a surfeit of politics explained the city's inability to clamp down on rampant violation of sewage treatment protocols. When Coluccio raised the issue with upstate engineers, "they would say that these sewage treatment plants are 50, 100, 125 miles away from City Hall, and we don't get the support we need. So they're crying poverty, we can't do anything about it."[29] In raising the issue of excess sewage discharges, Coluccio and his colleagues directly implicated the operations of their own department.[30] The DEP operated several of the largest sewage treatment plants in the watersheds, many of which were egregious violators of their state permits. As DEP Commissioner Harvey Schultz acknowledged in the late 1980s, the agency's substandard operation of its own treatment plants constrained its ability to pursue legal action against other violators: "It is not something DEP has done in the past, because I'm told that DEP felt that since our own plants . . . have significant non-compliance problems, it's a strategy that could backfire."[31]

A dearth of accurate water quality information also undermined efforts to protect Catskill watersheds. When Coluccio began to work with Gerald Iwan, the DEP's top water quality official, in the mid-1980s, he was asked to analyze water quality reports that detailed stream conditions in the watersheds. Every couple of weeks, upstate engineers would send handwritten reports on oversize ledger paper to their colleagues in New York for review. After poring through the reports for a few months, Coluccio became suspicious. "After a while, I said, 'You know, something's funny about this data; it looks like the data are too consistent to be real.' Because the environment doesn't work that way." Coluccio discovered that the employee responsible for producing the report simply duplicated the same numbers time and time again, and had apparently been doing so for several years.[32] The incident symbolized the city's static conception of the Catskills; despite the surge of development in the watersheds, it assumed that the region would provide pure mountain water for decades to come, allowing New York to maintain one of the country's last unfiltered municipal water systems. Bureaucratic inertia appeared to be more powerful than the new enthusiasm for ecology.

Coluccio and his colleagues struggled to elevate water quality to the top of the agency's agenda. Joseph Conway, a deputy commissioner committed to tackling the issue, created a water quality council, and tapped Coluccio to draft new watershed policies and regulations. With Conway's backing, Coluccio felt confident

New York could improve water conditions. Upstate staff appointed engineer Roy Frederickson to implement the new policies. But Frederickson avoided Coluccio's phone calls. It soon emerged that Frederickson had sought the position to promote his real estate interests in the Catskills. He used the information gleaned on the job to arrange sweetheart deals for land development in the buffer areas surrounding the city's reservoirs. These internal divisions undermined the agency's efforts to clean up its watersheds.[33]

Even as they struggled to overcome internal political conflicts, the DEP's water quality experts confronted a new threat from a proposed weakening of sewage treatment standards by the state Department of Environmental Conservation (DEC). The establishment of the DEC threatened to disrupt the traditionally cozy relationship between city water officials and state government. Established in 1970 and imbued with the era's ecological fervor (the legislation creating the agency was signed on the first Earth Day), the DEC was highly dedicated to its mission of protecting fish and wildlife. This mission frequently conflicted with the DEP's preferred management procedures. As the state agency responsible for promoting fishing, hunting, and other recreational activities, the DEC continuously pressed New York City to expand recreational access to its reservoirs and buffer lands, and to increase the volume of downstream releases below its dams. The agency also regulated sewage treatment plants throughout the state—inadequately, in the eyes of the city.[34]

By 1986, the DEP had begun to recognize the need for more aggressive protection of its watersheds, and all levels of the agency united to fight a preliminary directive from the DEC that permitted the sewage treatment plant in the Catskill Mountain town of Delhi to suspend chlorine treatment of its effluent. Treated sewage from the Delhi plant was discharged into the West Branch of the Delaware River, eventually flowing into the Cannonsville Reservoir. Although the Delhi plant was small, DEP officials worried about the precedent such a decision could establish. If the state suspended disinfection at plants throughout the city's watersheds, water quality could significantly deteriorate.

At its core, the Delhi case boiled down to two simple sets of parallel equations: city vs. state, and drinking water vs. recreation. State officials maintained that the Delhi plant should only have to comply with generic state standards on treatment plant operation, which did not require disinfection. Suspending chlorine treatment of effluent would improve the health of the stream's trout population. City officials countered that disinfection represented one of the major barriers to disease in the water system, and that the state failed to demonstrate that current disinfection methods harmed fish. In Vincent Coluccio's words, the state proposed to eliminate "a *proven*, major safeguard of public health in order to avoid a potential, low level risk."[35]

Despite securing the support of the state health department, which declared that suspending disinfection would significantly degrade water quality in the Cannonsville Reservoir, the city's odds of overturning the state's decision were uncertain because it fell to the DEC commissioner, not an outside court, to rule on New York's appeal. Eager to buttress its case with expert testimony from outside the state, New York City once again called on Abel Wolman. Still spry at ninety-four, Wolman consulted on water projects throughout the world and was widely regarded as one of the century's preeminent environmental engineers.[36]

In a rather bizarre turnaround, Wolman reprised his comments from the early 1950s about the vulnerability of New York City's watersheds. This time, instead of burying his report, the city put him on the witness stand. Likely experiencing one of the more delayed instances of vindication in his many decades of public service, Wolman happily obliged. In written testimony, he mocked the conventional portrayal of the system's watersheds as pristine: "This mythology persists up to now without justification." Although he largely reiterated the city's own arguments, Wolman did so with a bracing blend of authority and sarcasm, observing that "the serious issues of health and safety of the public water supply have been largely distinguished by their absence." After recounting recent outbreaks of waterborne disease in other states, he condemned the state for its failure to consider the gravity of the health issues at stake.[37]

Wolman's appearance at the hearing bolstered the city's case for disinfecting sewage before discharging effluent to water supply streams, and elevated the importance of watershed protection within the DEP. *Newsday* ran a lengthy profile of Wolman highlighting his role in the Delhi case.[38] The *Times* initially showed little interest in covering the story, but when Coluccio informed top editor Roger Starr that Wolman was involved, "he became very enthusiastic and wrote the editorial."[39] It is impossible to know whether media attention influenced the DEC's decision to decide in New York City's favor. The state ordered the continuation of disinfection, but insisted that the city pay for it.

By the mid-1980s, due to the restructuring of the water system's finances, the DEP could afford to pay for disinfection and a series of upgrades to its aging infrastructure. In 1984, the state legislature passed a bill creating two new entities: the New York City Municipal Water Finance Authority and the New York City Water Board. Instead of relying on city coffers to fund its substantial capital expenses, an arrangement that brought most capital construction to a grinding halt during the fiscal crisis, the new structure created a self-sustaining water system. The Water Board would set water and sewer rates high enough to pay for the operations of the DEP and to cover all the debt obligations incurred by the finance authority. As Albert Appleton, who served as DEP commissioner from 1989 to 1993, observed, "This sensible transfer of funding from general municipal revenues to user payments immediately ended the financial crisis for water and sewer infrastructure."[40] Complying with federal mandates and catching up on deferred maintenance proved expensive: from 1986 to 1993, rates increased 148 percent for water and sewer services.[41]

Infusion of additional revenue augured substantial changes in both the Catskill and Croton watersheds, including planned upgrades to DEP-operated sewage treatment plants, but critics questioned whether New York could make up for lost time. Frustrated by decades of inattention to safeguarding water quality, Coluccio's colleague Gerald Iwan blasted the agency upon his resignation in 1988. Iwan's caustic comments, which were featured in a *Newsday* exposé of New York's poor watershed management practices, put the DEP on the defensive and moved water quality higher up the public agenda.[42]

The *Newsday* investigation marked the emergence of watershed protection as a major public policy issue. Iwan and other DEP officials claimed that a third of the city's reservoir water was of borderline quality, and portrayed development in the watersheds as rampant, not only in Westchester County, but also west of the Hudson. Iwan described the engineering mind-set that he and Coluccio struggled against: "The city didn't put effort into water quality. We put effort into maintaining aqueducts and tunnels, not on protecting this valuable resource and the ecology which was giving us this resource. We didn't think there was a need for it. There was no industry, no development, no railroads. The water was good. It tested good."[43]

Although the flood of federal environmental legislation passed in the 1970s had little direct impact on New York's water system, by the 1980s the likelihood of future federal water-quality mandates convinced DEP leadership of the need to more vigorously protect drinking water sources. In 1986 Congress revised the Safe Drinking Water Act, requiring all cities served by surface water to filter their supplies unless they could meet a series of challenging water quality criteria. Watershed protection suddenly acquired a new urgency.[44] DEP Commissioner Harvey Schultz cautioned, "If effective preventive measures are not taken in time, the Catskill and Delaware systems will need the filtration that State Health and the USEPA are so eager to require."[45] Just as the 1970 Clean Air Act revolutionized emissions from tailpipes and power plants, the requirements of the Safe Drinking Water Act radically transformed New York's water system. Drawing its water from relatively pristine mountain watersheds no longer provided New York immunity from federal environmental oversight.

The DEP began to overhaul its watershed operations in the late 1980s. It initiated a water-monitoring program, established professional water quality labs in each watershed, and renovated city-operated sewage treatment plants.[46] These efforts signaled an end to the agency's traditionally cavalier attitude toward watershed protection, but were financial baby steps relative to the fiscal disaster New York City would suffer if the federal government ordered it to filter its Catskill and Delaware systems. Despite substantial increases in overall capital spending by the city and in the DEP's budget in the late 1980s, few additional funds were allocated for watershed protection.[47] Coluccio and Iwan had spent years in the political wilderness making their case; now it was Schultz's turn to play the role of Cassandra.

Increasing criticism of the DEP's lackadaisical approach to protecting its watersheds from environmentalists, the press, and the state and federal governments revealed a consummate irony: New York City's biggest mistake in managing its water system may have been a failure to use its political influence in the Catskills to restrict development.[48] A DEP attorney hinted that this failure was rooted in the agency's desire to maintain good relations with towns to which it paid substantial taxes. He argued that taking legal action to thwart development "would create a lot of problems in the long run" in the form of higher tax assessments.[49]

Tax disputes were, in fact, one of the major sources of tension between New York and Catskill communities. State law required New York to pay property taxes on all its extraterritorial holdings. In so-called impoundment communities (towns with city dams and reservoirs) New York often paid the bulk of local taxes. Financial reliance on New York City was both a blessing and a curse. As Eric Greenfield, who served as mayor of Delhi and played a crucial role in watershed protection negotiations, observed in the mid-1990s, "They're providing some communities with an incredible amount of taxes . . . one only has to look over the hill at the Downsville School district, a very small school district that's prospered primarily as a result of New York City tax money."[50]

New York's refusal to accept the fundamental premise of local assessors—that its waterworks properties were substantially more valuable than the homes and businesses that dotted the Catskills—unnerved Catskill officials responsible for providing services to watershed residents. For years, the city stamped its tax checks "Paid under Protest" to convey what it perceived as illegitimate assessments of its holdings.[51] It insisted that its property be classified as abandoned farmland, a designation that would have resulted in a substantial reduction in property taxes. Catskill officials countered that most city property consisted of rich bottomlands that had generated substantial tax revenues and that the city should pay at the same top rates as the landowners it had displaced. New York City routinely challenged the validity of its assessments, creating substantial financial uncertainty. Perry Shelton reflected on his experience as an official in the town of Tompkins, home to a portion of the Cannonsville Reservoir: "How do you run a town when you're being threatened with the loss of a quarter of the revenue from city-owned property when they own a quarter of the town?"[52]

New York City's perilous finances exacerbated the underlying fiscal uncertainty that plagued Catskill communities. Overwhelmed by the bills that piled up in the mid-1970s, New York delayed payment of its property taxes to watershed communities. Catskill residents joked about a potential default and the prospect of local communities assuming control over the dams and reservoirs. One journalist envisioned a booming entertainment district on the shores of the Ashokan Reservoir: "Disney Peninsula and Nuclear Light Village competed in the amusement park business" alongside casinos and convention centers.[53] Several years later, local photographer and writer Daniel Logan lamented the creeping development that threatened to mar "the majestic beauty of the Ashokan," but New York did not default, and no casinos rose from the reservoir's shores.[54]

Nevertheless, New York's fiscal collapse did create significant hardship for many watershed residents. Short of the revenue required to sustain services, the city had little choice but to practice financial triage; maintaining the bridges and roads that crossed and encircled its reservoirs quickly sank to the bottom of the to-do list. The inability to maintain reservoir infrastructure dramatically altered the rhythms of daily life for thousands of Catskill residents. In 1975 New York City closed the crumbling Traver Hollow Bridge, which linked residents on the south side of the Ashokan Reservoir to schools and shopping in the town of Boiceville. With the bridge out of service, residents were forced to drive around the reservoir to get to town. The bridge closing impaired fire and ambulance services to these residents, and "meant some West Shokan students had to board school buses an hour earlier in the morning and return home just before supper." Residents adapted the best they could, often leaving their cars parked on one side of the bridge and catching rides after walking across the bridge to the other side. They also vigorously challenged the bridge closing. In a June 1977 protest, residents hoisted signs linking their complaints to the energy crisis: "Conserve Energy! Open the Bridge." After Catskill residents sued, the city opened one lane of traffic. Ashokan-area residents rejoiced in 1980 when the state came to their rescue by opening a new multi-million-dollar crossing that stood right next to the original decaying bridge.[55]

Such conflicts stoked the resentment of watershed residents who had not forgotten the hardships that reservoir and dam construction had imposed on them and their forebears. From the perspective of New York City, waterworks construction had a finite beginning and end, but for watershed residents the reservoirs remained a fixture of daily life. The city's failure to maintain the roads around its reservoirs became a particularly contentious issue. Well into the 1990s, the poor condition of reservoir roads threatened to thwart larger efforts to reach a consensus over watershed protection. Perry Shelton waged a three-decade-long battle with New York City over its poor road maintenance. He recalled that "for 25 to 30 years they did *nothing* to them roads . . . and the potholes—you could hardly drive down that road and not hit a pothole."[56]

Reservoir Releases

In addition to sparring with the city over road and bridge maintenance, watershed residents challenged the recreational and ecological impact of water releases from Catskill reservoirs. In 1950, with New York City busy at work on three new reservoirs in the Catskill Mountains, a fisheries manager for the state Conservation Department envisioned the recreational benefits that the projects would bring to the region:

> It is not too much to imagine that the reservoirs will provide the City with the pure water it must have; that in the reservoirs there will be a balanced and properly managed fish population awaiting the angler; that the streams feeding into these reservoirs will be preserved as trout streams by the erection of barrier dams and that the streams below the reservoir will be guaranteed a constant flow of good water; that the lands around these reservoirs will be managed forests, with food and cover for wildlife and an opportunity for the hunter, campsites for the camper, trails for the hiker, picnic areas for the city family with just a few hours off.[57]

By the 1970s, this optimistic multiple-use recreational vision lay in tatters. Largely ineffective at regulating development and operating sewage treatment plants, the city excelled at a much more fundamental task: keeping people off its property. With the exception of fishing, which was permitted at city reservoirs under state law, the DEP prohibited almost all other activities on its lands and waters, including hiking and hunting.[58] The reservoirs and their buffer lands replaced farms and small towns with large inaccessible spaces that provided life's most essential resource to New Yorkers, but amounted to dead zones for most Catskill residents.

The streams that meandered through the mountains and valleys paid no heed to the sharp distinction between reservoir buffer lands and private landholdings implied by the ubiquitous "New York City Property: No Trespassing" signs. Industrial, agricultural, and residential activities in the Catskills inevitably affected the quality of the city's water. Conversely, reservoir operations, specifically the pattern of downstream releases (the discharge of water stored in reservoirs back into streams), strongly influenced flows and fishing conditions on many of the region's prized fishing streams. The Cannonsville Reservoir, more than any watershed property, illustrated the complex (and at times parasitic) relationship between city and countryside.[59]

The Cannonsville release controversy had its roots in one stubborn fact: water quality in the reservoir was substantially poorer than in the other Delaware reservoirs, the Pepacton, the Rondout, and the Neversink. In making its case to build Cannonsville, the BWS framed the debate as a choice between the filthy Hudson and the pristine West Branch. Intent on securing more mountain water, the board conveniently overlooked the fact that not all mountain water was created equal. Viewing the Catskills as a water-making machine of sorts for New York City, it ignored (or perhaps simply failed to appreciate) stark differences in ownership patterns, soil types, and land uses between the Pepacton and Cannonsville watersheds. The mountains were not an undifferentiated mass, but a mosaic of natural and human activities.[60]

In the Pepacton watershed, the combination of natural and human factors worked to the city's advantage. New York City was only the most recent outsider to purchase land in the Catskills. With the establishment of the Catskill Forest Preserve in 1885, New York State had begun to acquire parcels throughout the region. Concentrated primarily on mountaintops and steep slopes, state holdings posed little threat to the farmers and other residents who lived in the valleys, because such lands could not be developed easily and were generally unsuitable for agriculture. Shortly after the state began to acquire land for the forest preserve, wealthy residents of New York City and other urban centers established private fishing clubs on several Catskill streams.[61]

Many Catskill residents embraced these developments. Ownership of land by the state and assorted fishing clubs safeguarded Catskill streams and rivers, improving fishing well downstream of protected areas. Fishing clubs that purchased land from local farmers typically allowed them to continue to till the land if they took steps to protect stream quality such as refraining from watering their cows in the stream. Eager for the cash such sales generated, farmers willingly agreed to these conditions.[62]

When New York City began to develop its Delaware reservoirs, it became the beneficiary of these pioneering conservation initiatives. The physical and economic landscape of the Pepacton watershed was ideal for producing clean water. The state controlled virtually all the mountaintop land where the headwaters of the East Branch originated, and private fishing clubs owned stretches of riparian land along the Beaverkill and other tributaries. Moreover, poor soil limited the extent of dairy farming in the watershed, reducing the volume of manure and agricultural chemicals emptying into local streams. As one local historian put it, "the only thing that'll grow is rock."[63] This confluence of factors made Pepacton's waters exceptionally clean. To this day, Pepacton remains the city's premier fishing reservoir, its healthy population of brown trout a testament to its purity.[64]

Conditions were quite different on the West Branch of the Delaware, which the BWS dammed to create the Cannonsville Reservoir. Due to high volumes of agricultural waste in the West Branch and its tributaries, Cannonsville's water was the worst of the six mountain reservoirs. Heavy phosphorous loading from manure and agricultural chemicals produced algae blooms in spring and summer, occasionally resulting in large fish kills. Unlike other reservoirs in the system, Cannonsville never cleansed itself after construction, remaining exceptionally high in nutrients well into the 1980s.[65]

Why was the board so confident that Cannonsville was the equal of the other reservoirs in the region? Simply put, city engineers misread the complex Catskill landscape. Unlike the watershed surrounding Pepacton, which lay only fifteen miles away, Cannonsville's watershed was a productive agricultural community. The reservoir's anticipated effect on dairy farming in the watershed had figured prominently in the state hearings about Cannonsville. Watershed residents claimed that the reservoir would decimate the industry, the economic mainstay of Delaware County; city attorneys countered that dairy farming in the Catskills had already begun to wane because of larger market forces.[66]

The debate between tapping the Hudson and continuing to draw water from mountain streams overshadowed the potential flaws of the West Branch as a source of drinking water. Well aware that creameries and towns such as Delhi and Walton dumped raw and inadequately treated sewage into the West Branch, Catskill residents argued that the river was too polluted to supply New York.[67] The state commission brushed aside their warnings, focusing almost exclusively on the unsuitability of the Hudson. The denunciations of the Hudson by a parade of witnesses for New York City even convinced a local newspaper, which editorialized, "Certainly, testimony at Delhi last week tends to rule out the Hudson River as a permanent source of water for New York City, in favor of the more economical and purer source in the Catskills."[68] The West Branch, a font of purity compared to the Hudson, got a free pass.

Other factors compounded Cannonsville's pollution problems. The state owned little land in the watershed because it concentrated its purchases within the Catskill Park, which encompassed Pepacton, but not Cannonsville.[69] A dearth of protected land, combined with phosphorous from four sewage treatment plants that lined the West Branch, contributed to elevated nutrient levels.[70] However, the most significant threat to the reservoir's water quality was the concentration of dairy farms, which occupied over 20 percent of Cannonsville's watershed. The inability to control pollution from these farms led to "a severe eutrophication problem in the Cannonsville Reservoir."[71]

The vast disparity in water quality among the Delaware reservoirs dictated a skewed approach to their operation. Unwilling to send low-quality supplies to New York, officials frequently suspended diversions from Cannonsville. These suspensions typically lasted a few weeks. However, when nutrient loading caused enormous algae blooms, New York City sometimes went months without drawing on Cannonsville, relying instead on substantial diversions from the Pepacton and Neversink Reservoirs to meet its water needs.[72] The Supreme Court decree mandating minimum flows on the Delaware River did not specify the terms of compliance; the city could release an equivalent volume of water from its Delaware reservoirs, or concentrate its releases from one reservoir. It chose to make virtually all of its required releases from Cannonsville, sharply reducing the volume of water it returned to the Neversink River and the East Branch of the Delaware below Pepacton. As state biologists observed, New York used "the best water for drinking purposes and the worst water for meeting the Supreme Court order on flows."[73]

The changes in the downstream release regime following completion of Cannonsville radically altered the ecology and recreational patterns in the Upper Delaware basin. Trout fishing on the East Branch "took a nosedive as the releases from the Pepacton were greatly curtailed" following the completion of Cannonsville.[74] Substantial diversions and paltry downstream releases turned the Neversink River and the East Branch into sluggish streams most summers.[75] Fishermen now flocked to where the water was—below the Cannonsville Dam on the West Branch. Practically overnight, courtesy of the large volumes of cold water released from the Cannonsville Reservoir, the West Branch became one of the better trout streams in the state.

The releases from Cannonsville also transformed the upper reaches of the main stem of the Delaware River. Releases from Pepacton had marginally improved trout fishing in the Delaware by raising water temperatures, but because Pepacton was more than thirty miles from the beginning of the main stem, its influence was fairly limited. Cannonsville was located only eighteen miles from Hancock, where the East and West Branches joined to form the Delaware. When New York City released Cannonsville's frigid waters, it cooled the Delaware River, greatly extending the range of trout fishing territory downstream of the dam.[76]

Well aware of these general trends, the fishermen of the Catskills were nonetheless confounded by the inconsistent nature of the releases. State biologists documented erratic releases below Cannonsville "sometimes ranging from 20 cfs [cubic feet per second] to 1200 cfs within a 24-hour period, [which] destroyed the existing warm-water fishery and periodically made the river unfishable."[77] These abrupt changes in stream conditions eliminated the bass fishery and put trout populations under enormous pressure. Fishing the West Branch was a guessing game, glorious one day, impossible the next. As one report commissioned by the state observed, the downstream flow requirement "was not, it should be noted, included to protect the rights of those on the Neversink, East Branch Delaware and West Branch Delaware between the reservoirs and the main stem of the Delaware."[78] Concerned only with meeting the legal requirement regarding stream flows, New York City ignored the powerful effects of its release regimen on those closer to home.

Four decades after Cannonsville was completed, Catskill guide and conservation-ist Jim Serio summed up the problem: "When the Supreme Court ruling came down in '54, no one knew what ecology was."[79] But as the basic outlines of river ecology emerged, anglers identified the city's water practices as the crucial deter-minant of conditions on the Delaware and its branches.

Watershed residents grew increasingly frustrated with New York City because it refused to acknowledge that altering the release regimen to benefit the rivers would not compromise its water supply. The region's fishermen were convinced that the reservoirs could serve multiple uses. As one Catskill conservation orga-nization put it, "because of the New York City reservoir management policy, we are needlessly being deprived of some of the finest fishing imaginable."[80] The vast gap between the world-class potential of Catskill streams and the reality of erratic, sometimes impossible, fishing conditions inspired anglers to unite to challenge New York's release policies.

Although the Supreme Court decree language regarding stream flows applied only to the Delaware reservoirs, the operation of New York City's dams and reser-voirs changed fishing in other Catskill streams, particularly Esopus Creek. Con-struction of the Ashokan and Schoharie Reservoirs transformed conditions on the Esopus. Rainbow trout feasted on the abundant food in the Ashokan Reservoir, fattening up before returning to the creek to spawn in spring, when they presented an attractive catch to the anglers who lined the banks of the Esopus.[81] These trout fought heavy currents on their way upstream because of the regular release of wa-ter from the Schoharie Reservoir. Water from the Schoharie entered the eighteen-mile Shandaken Tunnel (known as the Portal by most Catskill fishermen), which connected the reservoir to Esopus Creek. The creek served as a natural conduit, conveying water from the Schoharie to the Ashokan, thereby saving New York the expense of constructing an additional ten miles of tunnel.[82] In theory, this arrangement also benefited anglers because it ensured an almost constant rush of cool reservoir water into a hallowed trout stream.

The reality proved far more complex. Inconsistent operation of the Portal and occasional turbidity in the Schoharie made angling on the Esopus an enormously frustrating experience, even to the many fishermen who had spent years master-ing its irregularities. As Paul O'Neil, part of a large contingent of New York City anglers who spent much of spring and summer haunting Catskill rivers, attested of the fisherman drawn to this fickle stream, "The Esopus . . . can drive him out of his mind."[83] The essential paradox of the Esopus—loaded with sizable trout, yet maddeningly unpredictable—attracted a devoted core of followers; O'Neil, for one, refused to fish any other Catskill stream. A man who thrived on the subtle give and take of fishing, the imperative to continually adapt to changing stream

Fig. 12. Paul O'Neil fishing Esopus Creek, 1964. This photo of O'Neil accompanied his May 1964 *Life* magazine article on the joys and frustrations of fishing the Esopus Creek. (Getty Images)

conditions, O'Neil confronted a city whose single-minded focus confounded even the most attentive of anglers: "The water department, naturally, is more interested in shooting 500 million gallons a day toward the Bronx than in an angler stranded at streamside by raging currents."[84] Always eager for a challenge, he nonetheless rued that decades after New York City reengineered the Esopus, its potential as a fishery remained unrealized.

It was on the Esopus that the conflict between fishermen and New York City came to a head. In October 1974, the city abruptly closed the Portal, cutting off the flow of water from the Shandaken Tunnel into Esopus Creek. The sudden shut-off did not give trout time to seek out deep pools where they could survive until the reopening of the Portal. Within hours, eager fishermen became unwilling undertakers, collecting hundreds of dead fish from several locations along the stream.[85] State officials estimated that over ten thousand rainbow trout perished, in addition to hundreds of walleyes, perch, and other species.[86]

The fish kill strengthened the bonds between Catskill fishermen, canoeists, river lovers, and the DEC. Well aware of the potentially catastrophic effects of a precipitous decrease in stream flow on aquatic life, state officials had secured an oral commitment from the city to gradually close the tunnel when it needed to perform inspections or repairs.[87] Embarrassed by New York's blatant disregard for its authority, the DEC considered filing suit. While not opposed to seeking a legal remedy, Catskill fishermen insisted that a change in release practices, not financial compensation, was their goal.[88] As Frederick Faerber, president of the Ulster County Sportsmen Federation, put it, "This city's got to be stopped. They can't continue to do this to us. New York City controls the water and don't give a damn about the fish."[89]

The release controversy exposed the blurry line between nature and artifice.[90] As the 1974 fish kill made abundantly clear, stream ecology depended on the proper application of technology; optimal functioning of the Esopus in turn required a delicate hand on the city's release mechanism. Well aware that New York controlled conditions on the creek, Catskill fishermen nonetheless experienced the Esopus as a natural site, a place where their minds and bodies would interact with those of other species. From New York City's perspective, the river was, to use historian Richard White's phrase, an organic machine, whose express purpose was to collect and convey water for its citizens.[91] This simplification of nature, in which the goal is to maximize the yield of a particular product—in this case high-quality water—is a hallmark of the pre-ecological state.[92] The challenge, as DEC Commissioner Peter Berle suggested, was to bring the reservoirs into the age of ecology. New York's reservoir system, he observed, "was developed over a period of more than a century during which very little was known about the ecological needs or values of the streams or rivers where the improvements were placed."[93]

Disparate conditions on the branches of the Delaware River underscored the critical role of reservoir release practices. Less than a year after the Esopus incident, another fish kill occurred, on the East Branch of the Delaware River. Even as substantial flows on the West Branch caused the Cannonsville Reservoir to spill, low flows on the East Branch left "exposed rocks baking in the sun with fish gasping for nonexistent oxygen."[94] New York's failure to release water from the Pepacton Reservoir led to a spike in stream temperatures, killing hundreds of trout.[95] Fish kills caused by extreme weather conditions had occurred in the Catskills before New York constructed its reservoirs. But once the city dammed the rivers and built massive reservoirs, Catskill residents and anglers charged it with the responsibility to intelligently regulate the region's waters. In manipulating nature, New York City soon became its guardian, just as the Army Corps of Engineers did on the Mississippi River in the twentieth century.[96]

State officials had traditionally endorsed Gotham's expansionist water supply policies. The DEC's strong resistance to the city's release practices marked a new era in which the state government increasingly pressured New York to operate its water system in a more ecologically responsible manner. While the 1954 decree required New York to maintain a minimum stream flow for multiple uses downstream of the reservoirs, the target flow rate, acknowledged the Delaware River master, was "a negotiated number that is hydrologically meaningless."[97] The DEC's insistence that New York City adopt a more nuanced release regime reflected the meteoric rise of ecological thinking and the attempt by state government to apply these new ecological concepts in the service of fish and wildlife.[98]

The dispute also revealed splits among the public itself. Both the city and the state claimed to be acting in the public interest, New York on behalf of water consumers, the state on behalf of fish, fishermen, canoeists, and other river users. The emergence of a capable state environmental regulator and a blossoming ecological consciousness altered the political calculus. Instead of opposing government interference, Catskill anglers welcomed it, recognizing that their best hope of changing release practices lay in forming a partnership with state government. The embrace of state government by Catskill residents appeared to be at odds with the stubbornly independent attitude exhibited by many mountain residents. But, as sociologists Catherine McNicol Stock and Robert Johnston have argued, rural Americans have alternately resisted and embraced government intervention.[99] In 1976, one group of watershed residents joined with urban anglers and state officials to lobby for changes in New York City's release policies. The following year, others challenged a state proposal to establish a regional agency to coordinate and regulate land use. Catskill farmer Einar Eklund critiqued the implicit message behind regional planning—that residents did not properly care for the land: "These new schemes to preserve the Catskills, proposed by the 'land planners and ecology advocates' fail to consider that local people operating in a free society have preserved this area for 200 years and will continue to do so."[100]

City and suburban anglers also found themselves pulled in two directions: they enjoyed fishing the reservoirs, but depended on New York City water to meet their daily needs. Mark Milani lived in a community that tapped into the city's water supply. An avid angler of the streams and reservoirs of Westchester County, he insisted that the two interests were compatible: "As a Yonkers resident . . . I benefit from the New York City watershed system. . . . The city of New York can have water and a great fishery can remain one also."[101]

These suburban anglers joined with Catskill fishermen to lobby state conservation officials to pressure the city to reform its release practices. For the first time, sportsmen from across the region banded together to form an umbrella group, known as Catskill Waters, to advocate for changes in the release policy.[102] The organization worked closely with officials from the DEC and the state attorney general's office to secure a written commitment from the city regarding new release procedures. Catskill Waters endorsed an alternate release plan drafted by the DEC that promised higher releases on the East Branch and Neversink Rivers and stabilization of flows from the Shandaken Tunnel.[103]

The city's refusal to implement test releases intended to generate data required to fine tune the proposed plan prompted state conservation officials to call for aggressive action against New York. DEC Commissioner Ogden Reid dubbed the existing release procedures a "desecration." Shortly after his resignation as commissioner in early 1976, Reid showed up unannounced at a Catskill Waters fund-raiser, where he received a standing ovation.[104] By that spring, anglers and their allies in Albany had abandoned all hope of reaching a deal with New York City. Their new goal was to enact state legislation to revamp downstream release practices.

Catskill Waters and the DEC each played key roles in the fight to regulate downstream reservoir releases. Both emphasized that the legislation would not result in less water for New York City; a more thoughtful release policy, they argued, would meet the needs of city dwellers, anglers, and environmental quality.[105] DEC support came principally from the organization's biologists and engineers, who chafed at what they perceived as New York's unbridled arrogance. Biologist Doug Sheppard secured office space for Catskill Waters president John Hoeko and reviewed the legislation that Hoeko drafted with the help of legislative aides. Bolstered by the DEC's empirical studies, Catskill Waters sought to appeal to legislators' emotions and to exploit upstate resentment of the city's prerogatives. Conscious of the need to portray the region's fishermen as moderate environmental reformers, Hoeko noted that the organization did not demand the removal of the dams, but simply that New York City bequeath "a legacy of healthy rivers to our children and their children."[106] Hoeko's deputy, Frank Mele, eagerly played up the coincidence of the legislative battle and the nation's bicentennial, accusing the city of "reverting to the same sort of dictatorial and oppressive rule which ignited the American Revolution." He concluded a May 1976 address with a rhetorical flourish likely to resonate in Albany, asking, "Are we Americans? Or subjects of New York City?"[107]

As May gave way to June, the answer remained unclear. Hoeko, a twenty-six-year-old construction worker with no prior lobbying experience, relocated to Albany, shadowing lawmakers as the legislative session drew to a close. Enraged by New York's unwillingness to seriously negotiate with Catskill Waters—"They would just say unequivocally [*sic*], unilaterally, 'No!' Just no to anything upstate"— Hoeko worked tirelessly to persuade lawmakers to rein in the city.[108] With passage of the bill in the rural-dominated state Senate assured, he focused his energy on the Assembly, where New York City traditionally enjoyed great sway. Three months after Saul Steinberg's classic *New Yorker* cover depicted New York City as the center of the world, and only days before the legislative session ended, Albany claimed center stage when the Assembly, by a single vote, passed the bill to regulate releases.[109]

A combination of fortuitous circumstances and long-term trends helped secure enactment of the release bill. Two state representatives worked tirelessly for the deal, resisting the efforts of New York City lawmakers to kill it. But those who followed the winding course of the legislation identified John Hoeko as the critical factor.[110] A thoughtful man who spent much of his spare time tying flies and fishing, Hoeko keenly understood the value of persistence. His strategy was simple: he constantly hounded lawmakers, forcing them to "eat, breathe, and drink me every day."[111] His doggedness paid off because it coincided with two emerging trends: the decreasing clout of New York City, and the increasing public receptiveness to nature preservation. Reeling from its fiscal crisis, the city found its leverage on the wane in Albany. Lawmakers there responded to their constituents' desire for expanded recreational opportunities, and displayed a growing sensitivity to ecology. The previous year, legislators had passed the State Environmental Quality Review Act (SEQRA), which mandated environmental impact statements and public participation for large development projects.[112] The environmental movement, with its emphasis on mitigating the effects of human action, had arrived. When it came wrapped in a "home rule" package, as it did in the case of the release bill, it appealed to a diverse group of politicians. The result was a seminal moment in the city's water history. As angler Austin McK. Francis observed, "for the first time since 1907 . . . New York City had made a formal accommodation to the rivers affected by its dams."[113]

The new release procedures revolutionized fishing in the Catskills. A few months after the implementation of the new releases on the Neversink, veteran angler Michael Longuil thanked the DEC commissioner for the improved fishing conditions: "Having fished this fine river the better part of my forty-five years, and having had to cope with a veritable trickle during the past few years, you have given people such as myself a very pleasant and relaxing summer."[114] A state task force shared Longuil's optimism, noting that the additional releases significantly increased habitat for trout and other cold-water species, and greatly reduced the frequency of dangerous temperature spikes.[115] More consistent releases revitalized trout fishing on the East Branch and Neversink, greatly expanding the length of stream habitat suitable for cold-water species.[116] However, most fishermen could not resist the lure of the West Branch of the Delaware River. The new releases moderated high summertime temperatures on the West Branch, eliminating the torrents that had often made fishing the stream impossible.[117] The result was perhaps the finest trout stream in the state. Catskill angler Jim Capossela humorously captured the magnetic draw of the river: "Now I know why God created the West Branch of the Delaware: to keep people away from the places that I like to fish by myself."[118]

Despite the marked improvements in fishing, it soon became clear to most Catskill residents struggling to reclaim their watercourses that they had won a significant battle, not the war itself. The complex political ecology of the city's water system effectively guaranteed that the release dispute would endure for the foreseeable future. Conveniently overlooking New York City's staunch resistance to the release bill, recently installed DEP Commissioner Francis McArdle suggested to Mayor Koch in 1978 that improvements on the upper Delaware streams represented "a major story opportunity" for the beleaguered metropolis.[119] By the 1980s, even the most optimistic civil servant would not dare make such a claim.

Political Changes, Political Stasis

The most significant development of the era was the newfound enthusiasm for regulating human use of the natural world. The willingness of government to protect ecosystems and individual species contained a powerful paradox: the wave of environmental legislation passed in the 1970s on the federal and state levels also subjected government actions to unprecedented levels of scrutiny. However, as the release controversy illustrated, environmental reformers struggled to overcome the legacy of government decisions made with little regard for aquatic life or recreational benefits.[120] In the absence of strong support from state government, Catskill residents lacked the political clout to enact meaningful reforms. The establishment of the Department of Environmental Conservation in 1970 changed the political dynamic.[121]

The new political arrangements—characterized by interactions among various levels of government and between the private and public sectors—ensured a more balanced approach to managing environmental resources. For much of the twentieth century, the city enjoyed substantial autonomy in building and operating its water network, exercising significant extraterritorial power in the process.[122] One advocate for improved fishing releases recalled a conversation he had in the 1990s with a longtime agency engineer, who told him, "in the '40s and '50s, New York City hired the best lawyers in the world, and we got the best deal for New York City, and we screwed upstate New York, and there's nothing you're ever going to do about it."[123]

The engineer failed to recognize that the city's near total control of watershed politics had begun to erode in the 1970s. Limited knowledge of ecology and reluctance on the part of state government to check environmental excesses blunted the capacity of the American political system to rein in New York's power.[124] Conscious of the need to balance urban and rural demands, the DEC recognized that the city had almost entirely disregarded the desires of Catskill residents and the ecological requirements of the rivers it tapped. When New York City refused to acknowledge that the environmental era had arrived, state officials joined with activists to force it to protect aquatic life, fishermen, and the upstate economy.[125]

By the early 1990s, however, the tensions of the previous two decades had given way to full-fledged rage on the part of most Catskill residents. New questions had emerged: Would the ties that had bound the watershed and New York City for so long finally rupture? And what would happen if they did?

CHAPTER SIX

The Rise of Watershed Management

When he became commissioner of New York City's Department of Environmental Protection in January 1990, Albert Appleton felt besieged. Albany and Washington had begun to scrutinize the operation of the city's water system, and neither liked what it saw. Federal law required the city to end within two years its longtime practice of dumping sewage sludge at sea. The state threatened to prohibit new sewer connections (effectively ceasing new building construction) in lower Manhattan and part of Brooklyn if the city did not reduce the volume of sewage flowing to Brooklyn's Newtown Creek facility, its largest wastewater treatment plant.[1] Most worrisome was the possibility that Washington would require New York to filter its water supply to comply with new federal water quality regulations. Constructing a plant to filter the billion-plus gallons a day the city drew from the Catskill Mountains was a daunting fiscal and logistical prospect. With a price tag of $6 billion to $8 billion (more than 20 percent of the annual municipal budget at the time), plus an additional $300 million in annual operating expenses, New York simply could not afford the filtration plant.[2] Water and sewer rates had been on the rise for years; the substantial increases required to pay for a massive filtration plant would likely provoke a rate revolt from landlords, businesses, and homeowners.[3]

There was an escape hatch, however. At the behest of New York and other cities with unfiltered supplies, the 1989 EPA regulation requiring filtration of all aboveground water supplies—the Surface Water Treatment Rule—contained a waiver provision exempting municipalities that met strict water quality criteria. To avoid the filtration requirement, the city would have to develop an aggressive watershed protection program to ensure the purity of its supply. The days of taking tentative steps to protect the water supply were over.[4]

Instead of working to expand its supply, as it had since the 1830s, the city concentrated primarily on protecting and preserving the water it already controlled.

After decades of expert reports recommending universal metering and other steps to curb New York's seemingly insatiable thirst, in the mid-1980s the city finally committed to monitoring and reducing its water consumption. In the 1990s, it began to make good on these promises. Regulatory pressure from other levels of government and the possibility of fiscal catastrophe propelled both water conservation and watershed protection. In each case, success hinged on the active cooperation of stakeholders outside of municipal government. In an era defined in large part by new technologies such as cellular phones and the Internet, both initiatives relied primarily on low-tech solutions to achieve results.

New York is one of a handful of large American cities that does not filter its surface water supply. Pioneered in Massachusetts in the late nineteenth century, filtration (through sand or mechanical devices) had within decades become the preferred purification method throughout the world. Even as industrialization and suburban sprawl degraded the quality of municipal supplies, filtration, and its sister treatment, chlorination, ensured that Americans could confidently drink the water that flowed from their kitchen faucets.[5] Most cities that chose not to filter—San Francisco, Seattle, and Portland, Oregon, are the classic cases—draw their supplies from federally protected watersheds with few permanent inhabitants.[6] No one could confuse the bedroom communities of the Croton watershed—with their cul de sacs, malls, and parking lots—with Yosemite, the source of San Francisco's water. DEP officials knew that without the benefit of filtration, the Croton water supply could not meet the stiffer federal standards. New York's failure to comply with the 1993 federal deadline to filter its Croton supply (the filtration plant is currently under construction and is scheduled to begin operating in 2013) was a product of a protracted dispute concerning the location of the filtration plant, not a rejection of the need for filtration.[7]

Overloaded sewage plants and second-home development notwithstanding, the west-of-Hudson watershed that supplied 90 percent of New York City's water bore little resemblance to the Croton system. Population densities remained low, few roads crisscrossed the region, and mountains, not malls, dominated the landscape. But even this relatively undeveloped region presented significant water quality challenges. Curbing contamination from point sources of pollution such as sewage treatment plants and factories posed few technical obstacles. The more difficult proposition was controlling nonpoint pollution, the waste that flowed from scattered farms, septic systems, and residences into the region's vast network of streams and rivers, eventually finding its way into the system's reservoirs. Could New York City, which had ignored deteriorating conditions in its watersheds for decades, really mount the kind of comprehensive watershed protection program required to secure a federal filtration waiver for its Catskill water supply?[8] Would more vigilant watershed protection stifle economic development in the Catskills?

More than thirty years after building its final reservoir, the city embarked on the most ambitious water supply challenge in its history: persuading Catskill residents to modify the way they farmed, developed their communities, and disposed of waste. This would be done to safeguard the water quality of a city many of them deeply resented. Dams and reservoirs may have taken years to build, but there was no blueprint for this job.

Regulating, Negotiating, and Litigating

From the city's perspective, pursuing a filtration waiver made ecological and economic sense. Given the choice between preventing water pollution and cleaning it up after the fact, prevention seemed the prudent option. Environmentalists who harshly criticized New York for ignoring mounting pollution in its watersheds agreed that filtration, while effective, was hardly infallible. A vigorous watershed protection program, they argued, was the best guarantee of public health. Environmental lawyer Robert F. Kennedy Jr., the water system's most vocal critic, observed that "the biggest threat to the watershed is pavement."[9] It was the enormous difference in cost between the two options that ultimately led New York City to apply for a waiver. Although the precise cost of watershed protection was impossible to calculate, it was clear to Commissioner Appleton that "the city could afford to spend a great deal of money on watershed protection and still come out vastly ahead of the game."[10]

Nothing underscored the city's nonchalance regarding its water supplies more than one oft-repeated fact: New York had not updated its watershed regulations since 1953. A potent symbol of the system's poor water management practices, the regulations were hopelessly outmoded; the outhouses and quaint pig farms they targeted had long since given way to septic systems and industrial agriculture. As a sign of its newfound commitment to source protection, New York released new draft watershed regulations in September 1990. The stringent regulations marked an entirely new conception of watershed protection in the Catskills. They aimed to protect water quality in the region's creeks and rivers by restricting a broad range of activities near watercourses, including farming and building. Because the regulations proposed an expansive definition of what constituted a watercourse, even landowners whose holdings were miles from major streams would face restrictions on logging, farming, and developing their parcels. The region's towns and villages, built astride the creeks and rivers that coursed through the valleys, feared that the regulations would severely constrain growth. Once again, New York City's need for water trumped the livelihood of Catskill residents.[11]

To many watershed residents, the regulations retained the familiar flavor of the imperial edicts New York had delivered to their grandparents when building the reservoir system: Get out now. Alan Rosa, a town supervisor from Middletown who helped spearhead resistance to the proposed rules, recalled the anxious months following the release of the regulations: "That was our main concern: that New York City would want to control population density and in the end virtually depopulate us."[12] According to Appleton, the regulations were intended to restrict the construction of second homes and other large-scale building.[13] However, those who read the fine print recognized that their potential repercussions went well beyond slowing the growth of vacation homes. They would severely constrain a wide range of activities, from expanding an existing village store, to building a parking lot, to planting corn.[14] Local officials throughout the mountain watersheds wondered how they would pay for the enhanced wastewater treatment mandated by the regulations. Farmers who had long resisted selling out to developers wondered if the end of Catskill dairying had finally arrived. The proposed regulations effectively banned all productive activity near the creeks and rivers that flowed through most Catskill farms.

By the spring of 1991, several months after the release of the draft regulations, two distinct opposition groups had emerged in the Catskills. The region's farmers, with help from the state Department of Agriculture, Catskill-area soil and water conservation officials, and professors and agricultural experts from Cornell University, had begun meeting with the DEP to devise an alternative set of agricultural regulations. A second group, composed primarily of town officials, took a more adversarial approach that envisioned active political and legal resistance to New York City's plans. Emotions were still raw when Catskill residents and officials gathered in the Delaware County village of Margaretville in March 1991 to discuss how to respond to the proposed regulations. One woman suggested blowing up the reservoirs. While most in attendance shared this sentiment, they set aside their anger in favor of forming an organized resistance movement.[15]

Ironically, the impetus for uniting watershed residents against the proposed regulations came not from the Catskills, but from lawyers based in Albany. As spring approached, attorneys representing Delaware County in unrelated legislation anxiously eyed the approaching deadline for public comments on the draft regulations. They convinced county leaders that individual watershed towns, many with fewer than a thousand residents, could not muster the resources required to do legal battle with New York City. They suggested convening a meeting of all watershed communities to craft a single response to the regulations.[16] Catskill residents quickly formed a united front. Despite the lack of a strong watershed identity—residents viewed themselves as members of individual communities, not as Catskill or watershed dwellers—they recognized their common interest in taking on the city. Anthony Bucca, a Greene County resident who was heavily involved in negotiating the watershed agreement with New York City, recalled the instantaneous sense of unity that emerged: "All of a sudden, we had a sudden identity thrust upon us. It was, 'Oh, you guys are in with us.'"[17]

By the time the meeting broke up, residents and officials had formed the organization that would challenge New York City's watershed protection plans in court. Bucca suggested calling the group OWEC, the Organization of Water Exporting Communities, but participants opted for the Coalition of Watershed Towns. Ken Markert, who convened the meeting in his capacity as Delaware County planning director, recalled that the recently concluded First Gulf War inspired the name. The prevailing definition of "coalition," eagerly promoted by the Bush administration, "meant that you got all these forces together that usually didn't work together and stomped the bad guys. That was very synonymous with what we were talking about."[18]

For the next few years, the coalition pursued a two-pronged strategy. Composed of elected officials from various watershed communities, the executive committee of the coalition included many members who had spent years battling the city over tax assessments and road repairs. These members used legal challenges to the regulations to prevent their enactment and also as leverage to obtain concessions on these long-simmering issues. At the same time, several towns joined New York City in a process called Whole Community Planning in which they agreed to take specific steps to improve water quality in exchange for relaxation of some of the more onerous regulations. Even as these negotiations proceeded, trust between the city and the coalition remained frayed. The DEP failed to follow through on simple tasks, such as road repairs, which would have demonstrated good faith. More fundamentally, Commissioner Appleton never accepted the need to seriously negotiate with the coalition. His opposition to residential development dovetailed with the desire of Catskill dairy operators to maintain their land in farms, but it ran headlong into the desire for economic growth in the towns and villages of the long-depressed region. Appleton failed to remove DEP employees who antagonized watershed officials, and he frequently belittled Catskill residents in public forums.[19]

While negotiations between the coalition and New York proceeded slowly, Catskill farmers and city officials made tremendous progress in forging a creative alternative to the original version of agricultural regulations. Only two of the hundred-plus pages of regulations dealt with agriculture, but they were the most controversial aspect of the plan. Drafted in New York City by, in the words of one state agricultural official, "people who don't know anything about agriculture," the new rules prohibited most farming activities within five hundred feet of a watercourse.[20] What might have proved a fairly innocuous restriction in some settings was much more ominous in the Catskills, where rivulets and streams meandered through virtually every farm. Catskill dairymen shared one attorney's analysis of the regulations: "They would have basically prohibited farming."[21]

Delaware County farmer Dave Taylor, whose grandfather and father had their farms seized by New York City to make way for waterworks construction, was determined not to suffer the same fate. He staked off the portion of his productive acreage that would be lost to farming under the proposed regulations—an astounding 45 percent—and challenged Commissioner Appleton to come have a look.[22] Appleton agreed. Gazing out at Taylor's shrunken farm, he realized the need for a new approach.[23] A chorus of outside voices echoed the call for a more flexible regulatory strategy. As county planner Ken Markert recalled, "Extension Service people, Soil and Water conservation district people [told] them they were totally out-to-lunch, the state agency people told them that. Cornell University told them that."[24]

When Albert Appleton arrived at the DEP several months before the release of the draft regulations in late 1990, he brought a new perspective to the agency. A former conservation chair of the New York City Audubon Society, he marked a sharp break with the engineering orientation that had long dominated the agency. An environmentalist and avid student of public policy, he believed that adopting a green agenda would allow the DEP to operate more efficiently. Instead of relying on technological solutions such as sewage treatment plants and reservoirs to ensure water quality, Appleton proposed a softer approach based on working with nature. Heeding the lessons of ecology was the right thing to do, he argued, but more important, it would save New York billions in infrastructure costs. From the perspective of Appleton and other cutting-edge environmentalists, conservation was no longer solely about preserving distant landscapes; it informed daily decisions in America's largest city.[25]

Appleton's new breed of environmentalism promised both risks and rewards for watershed residents. His desire to limit economic development and growth in the watershed and his tendency to assume a lecturing tone antagonized many residents and officials. Nevertheless, his commitment to ensuring an ecologically sustainable rural landscape offered hope that Catskill farmers and New York City could find common ground. Fiercely opposed to what he saw as the blight of unplanned second-home development encroaching on the watershed, Appleton recognized that it was in the city's interest to promote farming, not subdivisions.

The challenge was formidable: how could New York revamp the regulations to make them acceptable to Catskill farmers, while also sharply reducing the flow of pesticides and manure into streams that eventually emptied into the reservoirs? Appleton asked Dennis Rapp, deputy commissioner of the New York State

Department of Agriculture, for help. A Midwesterner with a farming background, Rapp recognized that consensus would prove elusive as long as the two sides continued to talk past one another. He insisted that the prerequisite to negotiation was education. Rapp organized a series of watershed agriculture forums in which city officials explained the financial burden associated with filtration and the corresponding need to clean up the watershed. In response, Catskill farmers described the severe financial burdens they faced; the watershed regulations, they argued, would render farming in the region completely impractical. But they also came to recognize that their tendency to load their crops with manure that washed off the land during storms introduced dangerous pathogens into the water system.[26]

By the end of the sessions, the two sides had established enough trust to form a joint task force charged with crafting an alternative regulatory approach that would improve water quality without compromising the viability of watershed farms. Pursuit of these seemingly contradictory goals led farmers and conservation experts to reconsider the fundamentals of Catskill agriculture. Like their counterparts in other regions, watershed farmers practiced a capital-intensive mode of farming that relied on machinery, grain feeding, and heavy application of pesticides. Reverting to a more traditional form of dairying based on grazing would allow farmers to slash costs by greatly reducing the use of pesticides and fertilizers. The task force's vision for the agriculture program mirrored the overarching filtration avoidance strategy. Rather than rely on expensive and unnecessary technology, farmers would carefully tend their herds and fields, in the process saving money and improving water quality.[27] Of course, changes of this magnitude raised thorny financial, political, and technical questions. Who would oversee and pay for the transformation of Catskill farms? What specific changes should individual farmers make?

For Catskill farmers, the most critical concern was maintaining their sense of autonomy over their own operations. If the ecological basis for what became known as the Whole Farm Program (WFP) was a return to a less capital-intensive mode of agriculture, the political basis was local control. Instead of the one-size-fits-all approach of the draft regulations, the task force argued that "a locally developed and administered program of best management practices, tailored farm-by-farm, with the voluntary cooperation of the farm operator, would be far more beneficial for both pollution control and the viability of the farm industry."[28] Local extension agents and soil and water conservation officials, not representatives of New York City, would work directly with Catskill farmers to overhaul their practices. In consultation with farmers, these officials would prepare Whole Farm Plans (WFP) to identify various "best management practices" that an individual farmer should adopt to enhance water quality and farm profitability. New York City would cover the costs of preparing the plan and reengineering the farm, from fence installation, to new plantings near streams, to systems to reduce barn-yard leakage. The plan called for a new administrative agency, the Watershed Agricultural Council (WAC), composed of local farmers, Catskill agricultural officials, and city and state representatives, to oversee the creation and implementation of the farm plans.

In little more than a year, New York City completely revamped its approach to curbing pollution from watershed farms. It abandoned traditional command-and-control-style regulations in favor of a voluntary, locally administered pollution prevention program. This decision implied a radical change in the way the DEP conceived of the watershed, or at least the portion of it devoted to agriculture. Rather than viewing the watershed as an undifferentiated whole, the DEP, through its support of the Whole Farm Program, indicated its willingness to align its farm protection efforts with the reality of the varied mountain landscape. Neighboring farms might adopt very different strategies to reduce the flow of manure and nutrients into streams.[29] In the 1950s, when New York condemned wide swaths of the Catskills to complete its reservoir system, one resident criticized the city's inability to appreciate the vast differences between farms that might have appeared similar: "In the city you move from one apartment to the next and maybe they are about the same. . . . There is as much difference between one farm and the next farm as there is between black and white."[30]

Four decades later, New York finally began to heed his message. It jettisoned the all-encompassing approach of the draft regulations in favor of a more tailored strategy. This shift reflected the recognition that, from the perspective of water quality, well-managed farms were preferable to new housing developments. It also indicated a more sophisticated appreciation of the political ecology of watershed protection. The DEP recognized that improving water quality within the watershed was a two-front war. More thorough treatment of point sources of pollution, especially the discharge from wastewater plants that flowed directly into Catskill streams, was critical, but the fight against water pollution also had to target nonpoint sources, such as agricultural waste. After prodding from outside parties, the agency learned that in the battle against nonpoint pollution, brute force and technological sophistication made poor tools; indeed, the martial metaphor itself was inapt. Reducing pollution from nonpoint sources would require the active cooperation of watershed residents. Securing this cooperation required New York to devise a more individualized approach to watershed protection that offered tangible benefits to participating landowners. After decades of viewing the watershed from on high—of, in James Scott's phrase, "seeing like a state"—New York began to incorporate the perspectives of those who lived in the Catskills. Preserving an unfiltered water system required a finely tuned appreciation of local variations.[31]

The push for watershed protection prompted both sides to reevaluate their traditional views of the region. The WFP negotiating process effectively separated the concerns of Catskill farmers from those of the larger community. However, with their way of life threatened, other Catskill residents set aside their local allegiances to unite as watershed dwellers. New York City moved in the opposite direction, parsing the ecological complexity of the mountains in much finer detail than it had ever attempted.

Environmental organizations and the federal Environmental Protection Agency did not share New York's enthusiasm for this collaborative mode of environmental policy. Activists and government officials were more comfortable with a traditional top-down regulatory approach to ensuring water quality, and were deeply skeptical of the voluntary nature of the WFP. Even if the program succeeded, they argued, it would not lead to substantial improvements in water quality unless the vast majority of farmers agreed to participate.[32] Yet for farmers, as for Catskill property owners more generally, the ability to opt out of watershed programs was a critical component to any deal. Commissioner Appleton struggled to chart a middle course between mandatory regulations on the one hand and complete voluntarism on the other. He and Richard Weidenbach of the Delaware County Soil and Water Conservation Service negotiated a compromise under which the program would go forward only if 85 percent of watershed farmers participated.[33] Achieving a high rate of participation in the Whole Farm Program, Appleton argued, would produce better results than a mandatory nonpoint pollution plan that attempted to regulate a broad range of activities. A mandatory plan would be impossible to enforce throughout the sprawling watershed.[34]

A creative blend of local control and big government funding, the Whole Farm Program ultimately succeeded because reducing pollution and maximizing farmers' income proved compatible. WAC built support for the program among Catskill farmers by carefully selecting pilot farms to participate, explaining the potential benefits, and positioning itself as an intermediary between the farmers and the city. WAC board chairman Fred Huneke recalled the recruitment process:

> The city said, if you join this program, then you will be exempt from city regulations . . . so that that was the carrot. So once we said that, and once we said that no matter what we do on your farm it's going to be paid for 100% and we're not going to destroy your business; we're going to help your business. So once they started to see that the 10 pilot farms were starting to work, then they just piled on. They really did. And it worked; it really did. But I think the key to that was the fact that we stand between the agricultural community and the regulatory community. That's why it works.[35]

A resounding 93 percent of watershed farmers elected to participate in the program. Although dairy farming remained an economic mainstay in parts of the watershed, it had been declining for decades. Farmers struggled to cover the expense of running their machines and feeding their cows. When the city agreed to finance the transition to a less capital-intensive form of agriculture—one that had potential to greatly reduce costs, and thereby increase profit margins—watershed farmers seized the offer. As Appleton noted, "Catskill farmers who had previously thought of the environment as something that forced them to spend their money to help others were now making money by becoming stewards of environmental resources, money that was helping them stay in farming."[36]

Within a few years, the changes were visible throughout Delaware County, the epicenter of watershed dairy farming. Farmers largely dispensed with row crops (used to feed cattle) on sloping hills; the cow manure used to fertilize such crops often ended up in nearby streams, degrading water quality. By grazing their cows and more selectively applying manure to their remaining crops, watershed farmers significantly reduced the volume of nutrients and pathogens emptying into Catskill streams.[37] The Davis family farm in the Delaware County community of Kortright exemplified the new approach. In addition to constructing a manure storage facility, the family planted more than sixteen acres of trees and shrubs along a tributary of the West Branch of the Delaware River to capture and absorb nutrients before they entered the stream. The Davises also installed a series of fences to keep cattle out of hydrologically sensitive areas.[38]

When multiplied across dozens of farms, these small changes significantly improved water quality in the Cannonsville Reservoir, and served as a model for effective control of nonpoint pollution.[39] By the end of the 1980s, federal and state regulators had successfully clamped down on point sources of pollution such as drainpipes, factory smokestacks, and the like. Nonpoint pollution posed a much greater challenge. In the Catskills, where the actions of thousands of individuals and businesses determined the health of landscapes and watercourses, strict regulation of a handful of highly polluting sites was unlikely to yield substantial water quality benefits. Legislators and regulators recognized the need for a different approach to tackle nonpoint pollution, but struggled to craft an adequate substitute. The Whole Farm Program, regarded as one of the most successful nonpoint pollution programs in the country, heralded a new way forward. Instead of treating residents as children who needed to be monitored—an impractical approach akin to having one babysitter for hordes of misbehaving youngsters— the WFP provided watershed farmers, the largest sources of nonpoint pollution in the watershed, a powerful financial incentive to curb discharges that could degrade water quality.[40]

Fig. 13. Moody Farm—before. Cows on this beef farm in the western Catskills used to feed along the stream during winter, adding to the pollution that ended up in the Cannonsville Reservoir. (Watershed Agricultural Council)

Fig. 14. Moody Farm—after. One of the most common best management practices paid for with watershed agricultural funds was the construction of barnyards, which dramatically reduced the volume of animal waste in area streams. (Watershed Agricultural Council)

A consensual mode of environmental governance emerged in the early 1990s in response to the limitations of scientific knowledge, the rise of the property rights movement, and the frustrations inherent in attempting to resolve environmental conflicts through the courts. On a variety of issues—watershed management, federal grazing policy, habitat conservation plans for endangered species—dueling parties agreed to negotiate rather than pursue their goals through the legal system. As political scientist Richard Andrews observed, "Broad-based stakeholder negotiations offered the hope of building workable compromises among legitimate conflicting interests, rather than continued trench warfare."[41] The WFP reflected the blending of scientific and local knowledge at the heart of this new approach. By including a wide range of stakeholders—property owners, local government officials, the local representatives of federal agricultural agencies, and New York City—the agricultural task force increased the likelihood of crafting a durable solution to the problem of farm-based contamination of the water supply.[42]

The WFP was also significant because it served as a model for the comprehensive watershed agreement reached in 1997. In addition to insisting on meaningful local input and the lure of financial incentives, the farm program relied on state government to act as an impartial broker, a crucial element in both sets of negotiations. In the case of the WFP, however, the state proved a reluctant partner. At Appleton's request, Dennis Rapp agreed to mediate between city and watershed interests. Rapp's superiors in the agriculture department tolerated his activities, but officials in Governor Mario Cuomo's office made it clear that "if you screw up, you're going to get hung out to dry."[43] Rapp skillfully managed the farm negotiations. His decision to start with an educational process laid the critical foundation of mutual understanding required to reach a settlement. Appleton, regarded as arrogant by many watershed leaders, nonetheless willingly acknowledged Rapp's critical role in crafting the WFP.[44] As a representative of the state government, Rapp was a logical intermediary. But it was his farming background and experience negotiating legislation—he helped write New York State's landmark 1988 Solid Waste Act—that enabled him to launch the constructive dialogue that eventually culminated in the WFP.

Progress on implementing the WFP appeared to carry over to the larger question of filtration. Establishment of the farm program, success in meeting an extended deadline for ceasing the dumping of sludge in the Atlantic Ocean, and political ties to high-level EPA officials earned New York a dramatic reprieve: hours before leaving office in January 1993, EPA Administrator William Reilly ruled that the city would not have to filter its water.[45] Reilly's decision was provisional; the federal government would continue to grant New York City filtration waivers only if it implemented a vigorous watershed protection program, including substantial land acquisition in the Catskills. City land purchases would relieve development pressure, safeguarding water quality for the long term.[46]

Soon after receiving its filtration waiver, New York City applied to the state for the right to purchase land in the watersheds, a fateful decision that shaped the course of all subsequent negotiations. Most Catskill residents deemed the filing of the application a declaration of war. The decision rankled on several levels. Residents worried that land acquisition would reduce their property values by placing large chunks of the region off-limits to development. But their greatest fear was a recurrence of their grandparents' experience—if New York could not purchase enough land to satisfy federal requirements, would it resort to eminent domain and condemn Catskill farms, homes, and businesses?[47]

The decision to pursue land acquisition derailed the ongoing negotiations between the city and individual Catskill towns. As Keith Porter, a Cornell scientist who helped broker some of these negotiations, recalled, the DEP's permit application "dissolved the trust and goodwill that had been developing over many, many months and many, many meetings."[48] The headlines of Catskill newspapers—" 'The Avenue of Attack That Will Kill Us': Condemnation Is New York City's Intent," "Robert Kennedy Says Condemnation Now!"—vividly conveyed the fear and enmity that now gripped the watershed.[49] Negotiations ground to a halt. The Coalition of Watershed Towns filed a lawsuit to block the state from granting the city land-acquisition permit. As 1994 dawned, and New York City welcomed Mayor Rudolph Giuliani to office, the future of its water system hung in the balance.

One Apartment at a Time

The impasse in watershed negotiations did not delay the DEP's efforts to substantially reduce water usage in the city. A few years before the agency shifted its stance on managing the quality of its supplies, it broke with decades of established policy by acknowledging the need to curb excessive water consumption. Issues of quantity, as well as quality, topped the DEP's agenda in the 1990s. Pressure from external forces, the fickleness of nature, and changing perspectives within the agency combined to produce a new approach to water conservation.

This shift had its roots in the events of the 1980s. After two droughts in five years, Mayor Koch made a momentous decision in 1986: New York City would meter all water delivered in the city, from bungalows in Brighton Beach to high-rise apartment buildings in the Bronx and Manhattan that housed more residents than most Catskill towns. Metering was the most important plank of what the mayor called his permanent water conservation program. The initiative had several components, including installing sonar leak detection to pinpoint waste in the distribution system, disconnecting abandoned buildings from water mains, launching a wide-ranging public education campaign with the slogan "Don't Drip New York Dry," and enacting new legislation to mandate the installation of low-flow fixtures in new and renovated buildings. The DEP's new water conservation unit launched a water-watcher calendar that featured art created by schoolchildren. One student drew a dystopian image of a parched city strewn with camels and cacti; "NEW YORK COULD BECOME A DESERT," she cautioned her fellow citizens.[50] An intergovernmental task force created by Mayor Koch urged the city to pursue an even more ambitious conservation strategy, which it estimated could cut water consumption by as much as 25 percent.[51]

The state government and environmentalists had begun to actively prod New York to accelerate its water conservation efforts. New regulations required the DEP to obtain a state permit to operate its Chelsea pumping plant, which it had put into operation during a severe drought in 1985. In protracted state administrative hearings that began in 1986 and continued for several years, environmentalists argued that more effective water conservation would obviate the need for the plant, which they claimed seriously damaged aquatic ecosystems. In response, the DEP eagerly touted its recent string of conservation accomplishments. Ironically, the ambitious goals set by the mayor's task force undermined the city's claim that it had made substantial progress in trimming water consumption. Using piquant language, state environmental lawyers rejected New York City's proposed conservation program, deeming it too little, too late. Arguing that "the city has no conservation ethic," a state attorney compared New York to "a junkie who, confronted by the evidence of its habit and offered a relatively painless means of withdrawal, first denies that its addiction is debilitating and then rejects the pleas of its family to go straight."[52]

Unable to satisfy the state or the environmental community, Commissioner Appleton elected to withdraw the city's permit application for the Chelsea plant in 1991. Appleton's environmental background predisposed him to embrace a more ambitious water conservation agenda. Shortly after taking office, he declared, "I don't agree that it is inevitable that we expand the water supply."[53] He also had a much more powerful incentive to reduce consumption than merely obtaining a permit for the Chelsea plant: the threat of a state construction moratorium in neighborhoods whose sewage treatment plants were overloaded. Excessive water consumption taxed the plants, rendering them unable to adequately treat the enormous volumes of sewage they received. New York responded to a possible construction freeze by prioritizing the installation of water meters in those neighborhoods whose wastes fed into the targeted plants. But this was only a stopgap measure.

Appleton had a much larger vision for bringing the water supply into the environmental era. Working with nature and curbing water consumption—the pillars of his new approach—promised significant benefits on many fronts. A substantial reduction in water usage would enable New York to sharply reduce flows entering sewage treatment plants, bringing them into compliance with their state permits and removing the threat of a construction moratorium. Reducing wastewater flows would in turn eliminate the need for additional sewage treatment facilities, saving the city at least several hundred million dollars and probably much more. Slashing water usage would also enable New York to weather a drought without relying on the Chelsea plant. Most significant for the long term, implementing these measures would also finally put an end to the perpetual search for new water supplies. One conservation expert noted the sharp departure from long-standing city water policy: "The pattern of building the supply to meet demand was officially broken, a fault line crossed."[54]

The DEP's urban environmental agenda eventually proved much more popular with New York City's power brokers than its vision of greening the countryside did with elected officials in the Catskills. Landlords had squashed all previous attempts at water conservation and opposed the plan to base charges on consumption as measured by the water meters being installed throughout the five boroughs. They feared that meters would result in substantially higher bills, and that rent-control regulations would prevent them from passing increased costs on to their tenants. Appleton convinced New York's real estate community that the capital expense of expanding the water supply would far exceed the cost of obtaining a "new" supply by reducing consumption. Investing in conservation would stabilize rates, not increase them. He also gained the support of contractors by exploiting a fortuitous coincidence: Francis McArdle, the former DEP commissioner, served as managing director of the General Contractors' Association. McArdle convinced his members that there was more money to be made on repairs to water mains and other conservation-oriented projects than on a single upstate water filtration plant. In a sign of how dramatically attitudes began to change, the impetus for New York's most successful water conservation initiative—its toilet rebate program—came not from municipal officials, but from a representative of the Plumbing Foundation, a nonprofit association of contractors, plumbing suppliers, and manufacturers.[55]

In the 1990s, New York embarked on the world's largest toilet replacement project to modernize one of its most outmoded private infrastructure systems—the millions of toilets inside homes and apartments.[56] A typical New York City toilet used between three and four gallons per flush; thousands of antiquated World War I–era pull-chain models consumed over five gallons per flush. Old toilets also had a tendency to leak. Accustomed to the cacophony of urban life, many New Yorkers appeared deaf to the sound of a running toilet. Renowned Catskill fisherman Art Flick remarked incredulously that he occasionally visited friends in

New York City, and "almost every time I find the same leaky faucet that was dribbling away our water two years ago." Installing low-flow toilets, which consumed only 1.6 gallons per flush, had the potential to dramatically reduce water consumption without compromising quality of life.[57]

New York City rebounded from decades of socioeconomic degradation during this period. Crime rates plummeted, immigrants flocked to the city, the economy expanded, and the Yankees reemerged as baseball's perennial powerhouse. Meanwhile, inside hundreds of thousands of apartments, the DEP unleashed a porcelain revolution. From 1994 to 1997, under a generous rebate program that cost almost $270 million, New York City replaced 1.3 million toilets in private residences. The effects were dramatic and, by conventional bureaucratic standards, lightning quick. As program director Warren Liebold observed, the massive scale of toilet replacement manifested itself almost immediately at wastewater treatment plants: "Toilets were being replaced at such a rate that you could look at the month-by-month dry weather flow averages at the Hunts Points and Wards Island plants and watch them go down. It was at that sort of scale."[58] Well aware that they would eventually have to pay for water on a metered basis and that the city would not again offer generous subsidies for toilet replacement, owners of large multifamily buildings were quick to join the program. They received $240 for the first toilet or showerhead replaced in an apartment and $150 for additional replacements. Required by law to replace at least 70 percent of the toilets in a building, they often chose to replace 90 percent. These buildings housed some of the poorest New Yorkers, people with large families who were often home during the day and used significant amounts of water. The results validated the high expectations for the program: water consumption in participating buildings decreased an average of 29 percent.[59]

Despite a booming economy and a rapidly expanding population, New York reversed a century and a half of inexorably increasing water consumption. In 1991, consumption averaged almost 1.5 billion gallons a day. Historical trends forecast consumption of 2 billion gallons by 2010.[60] Instead, the combination of universal metering, toilet replacements, an aggressive education program, and improved leak detection slashed consumption by almost 25 percent by 2003.[61] Because the city implemented several water conservation initiatives simultaneously, estimating the reduction in consumption attributable to a specific intervention is difficult. But Liebold identified metering as the keystone of the toilet rebate program and the larger water conservation initiative:

> The stuff that's aimed at the customer would not have gotten very far if we weren't metering, because why would you do it? I mean the apartment building owners would not have had a strong feeling about doing it one way

or the other, if not for the fact that even if they didn't have a meter in their building right then and there, they knew the writing was on the wall, and sooner or later they were going to have a meter, and God forbid, they'd end up being billed on a meter, they weren't going to turn down an opportunity to do what they could to reduce water use. So, you know, breaking out the effect of metering is difficult to do, but it's also fair to say that you'd be nowhere without it.[62]

With metering a sword of Damocles threatening to cut into their thin profit margins, owners of multifamily apartment buildings took long-overdue steps to curb water consumption. In addition to replacing toilets under the rebate program, owners began to emphasize leak prevention and water fixture maintenance; one landlord even offered building superintendents a free vacation for meeting consumption reduction targets.[63]

New York's remarkable success in cutting water consumption had much in common with its more heralded efforts at fighting crime in the 1990s.[64] Products of carefully crafted public policy, both initiatives changed the economic and physical landscape of the city. Steep reductions in violent crime prompted tourists to return to New York in droves and spurred development throughout the city, transforming neighborhoods in every borough.[65] In a sort of reverse image of the flurry of activity unleashed by the success in fighting crime, the water conservation initiative's chief accomplishment was precluding the construction of new water and sewer supply infrastructure. The tremendous savings, conservatively estimated at $4 billion to $5 billion, but potentially more than double that, explain why Mayor David Dinkins, presiding at a 1992 ceremony commemorating the 150th anniversary of the Croton Aqueduct, proudly announced the cancellation of new waterworks construction on the Hudson River and the shelving of plans to build additional sewage treatment plants in the city.[66] In sharp contrast to the wave of gentrification that forced many New Yorkers to leave their neighborhoods in search of less expensive housing, municipal government implemented its water conservation program without displacing tenants. The DEP phased in metered billing for apartment buildings and devised alternative arrangements for the subset of properties whose water bills threatened to increase substantially despite their best efforts to conserve.[67]

Albert Appleton did not hold office long enough to fully implement these water conservation initiatives. In November 1993, Rudolph Giuliani defeated David Dinkins in the race for mayor, and Appleton was out of a job. Giuliani replaced him with Marilyn Gelber, who came to the DEP from the city's planning department. Despite the significant progress Appleton made in his time as commissioner, he left Gelber an enormous challenge: reaching an agreement with Catskill residents on a watershed protection plan that would enable New York to avoid constructing an enormously expensive water filtration plant.

Watershed Protection Gets New Life

In early 1994, few DEP employees or Catskill residents held out much hope for the success of watershed negotiations. The Coalition of Watershed Towns filed a series of lawsuits to prevent New York City from implementing its watershed regulations or acquiring land in the watershed. City attorneys cautioned Commissioner Gelber to stop "fantasizing about an agreement with watershed communities"; the fate of the city's water supply, they insisted, now lay with the courts.[68]

Relying on favorable legal decisions to avoid the enormous expense of filtration made little sense to Gelber. Just as Dennis Rapp recognized that no agreement on watershed agriculture was possible without a process of mutual education, Gelber appreciated that resuscitating negotiations depended on improving the relationship between the city and watershed interests. Her status as an agency outsider and as the DEP's first female commissioner enhanced Gelber's ability to repair relations with Catskill residents. Leery of trusting anyone inside the DEP, she traveled alone to her first meeting in the watershed. For once, the roles were reversed: the coalition and its lawyers vastly outnumbered the lone representative from the big city. Unaware that offering watershed residents direct access to a DEP commissioner was unprecedented, Gelber urged those in attendance to call her directly.[69] She also embarked on a listening tour of the Catskills, meeting with key decision makers to get to know them personally and to better understand their concerns. According to Perry Shelton, a founding member of the Coalition of Watershed Towns, Gelber played the crucial role in repairing the city's relationship with the watershed: "It really changed considerably with Mrs. Gelber—Commissioner Gelber. She was just an altogether different person. We had some pointed exchanges with Appleton and some of the others early on. . . . Now we talk to each other. Mrs. Gelber . . . calls me up and talks to me. I call her up anytime I want to. She never refused to take a call."[70] The Catskill press also detected a change in New York's attitude. After Commissioner Gelber met with Delaware County officials in September 1994, a local newspaper dared to view the situation from Gotham's perspective: "Yet for the city to swallow the ultimate lump and place its commissioner at the bay of 19 supervisors was perhaps to say you're right. Chew us out. Get it over with and let's get this watershed program moving."[71]

With New York's filtration exemption up for review in 1996, Gelber needed evidence of demonstrable progress on the two most controversial issues—land acquisition and watershed regulations—to persuade regulators to extend the waiver.

Seeking to capitalize on improved relations, in the fall of 1994 she proposed a settlement, but the coalition rejected the deal because it lacked economic development funding to compensate watershed communities for more restrictive water quality regulations. The timing of the pact made little sense from the coalition's perspective because Republican George Pataki, who hailed from the Croton watershed, was locked in a tight race to be the state's next governor.[72] His vigorous support for the prerogatives of watershed residents helped seal Mayor Giuliani's decision to buck his party and support the reelection of Democratic governor Mario Cuomo. Giuliani feared that a Pataki win would render the construction of the filtration plant inevitable.[73]

Pataki's surprise victory changed the complexion of watershed negotiations. Accustomed to being ignored by state government, Catskill residents now had the most powerful man in Albany on their side. As a state senator, Pataki had twice supported a bill terminating New York City's right to regulate activities in the watershed.[74] In his new position as governor, he stood to benefit politically if he could end one of the most contentious battles in the rift between upstate and downstate. Pataki credited paralyzed actor Christopher Reeve, who had joined environmentalists in urging more aggressive watershed protection, for inspiring him to broker peace in the watershed.[75] Whatever his motives (Pataki was also a committed environmentalist, a factor that likely played into his decision), the governor's involvement proved decisive. In April 1995, after only three months in office, he ordered his counsel, Michael Finnegan, to bring together the city, the coalition, and the federal government to negotiate a settlement. At the first negotiating meeting, Pataki asked the participants to achieve the impossible: reach a deal within four months.[76]

A fortuitous set of circumstances laid the groundwork for a potential settlement. The clarifying effect of stringent federal regulations was the most important factor. Absent the threat of filtration, New York City had no compelling reason to enter into a cooperative agreement with watershed residents and officials. Although the federal government did not play a leading role in negotiating the details of a watershed pact, it loomed large over the whole process. Crafting an agreement that would pass muster with the EPA was an essential component of any settlement.

However, as the collapse of previous negotiating efforts and the coalition's lawsuits indicated, the likelihood of federal action alone was not enough to produce a deal. The tattered relationship between New York and watershed interests obscured the logic behind reaching a deal. Some in the Catskills questioned the impetus behind negotiating with New York. After all, the city, not rural residents, would bear the cost of building a filtration plant. New York would be less likely to insist on strict watershed regulations if it had the peace of mind that would come with a modern filtration plant. Coalition negotiators understood such sentiments, and shared them to some extent, but concluded that the alternative—fighting the city in court—offered little hope for success. Moreover, as coalition executive committee member Anthony Bucca observed, "The federal government and the state were not going to let the City go down the tubes. . . . So it really was a no-brainer. We had to try to make the best of it."[77]

What ultimately brought the coalition to the negotiating table was a change in government personnel. Sympathetic to New York City's position, the Cuomo administration did nothing to broker a comprehensive settlement. It believed that the city would ultimately prevail in the courts, and saw no need to compromise with rural interests. Without a neutral party to shepherd the negotiations, the prospect for a successful outcome appeared remote. The coalition's faith in the Pataki administration's ability to serve as an honest broker decisively altered the political calculus. By 1995, negotiating appeared a more prudent option than litigating. But as Daniel Ruzow, the coalition's lead attorney, recalled, it was the presence of both Pataki and DEP Commissioner Gelber that offered hope for a negotiated settlement: "The Governor made it clear to us . . . how important he saw resolving this issue was. . . . You had a key player respecting the rights of the upstate communities. Marilyn . . . conveyed the same perspective of respect for the local issues. So you had a good setting, finally."[78] For watershed residents, government was no longer a distant, unfriendly abstraction. It was a force embodied by actual people who were willing to listen to their concerns.

The leverage of city and state officials over watershed interests derived only partially from their formal powers; their ability to demonstrate a cultural affiliation with rural residents also proved critical. The importance of good interpersonal relations is underscored by the failure of previous administrations to form similar bonds, a failure that hampered efforts to negotiate a watershed agreement. The influence of Governor Pataki and Commissioner Gelber greatly improved the prospects for peace in the watershed. These powers of cultural affinity loom large when the parties involved represent disparate cultures and backgrounds.[79]

Despite the improved climate created by Pataki and Gelber, mistrust permeated the early stages of negotiations in the spring of 1995. Erin Crotty, who represented Governor Pataki throughout the negotiations, recalled that "in the first meetings that we convened, there wasn't a lot of eye contact." It soon became clear that the success of the negotiations hinged on the ability of state officials to act as an impartial broker whose sole interest was striking a deal acceptable to both sides. Crotty and other state officials prepared daily agendas and conferred with the parties away from the negotiating table to gauge their concerns and their willingness to compromise on particular issues. These informal conversations gave Crotty and her colleagues insight into how to structure the negotiations and how to approach sensitive topics.[80] Cognizant of the need to build momentum and establish trust between the two sides, state officials raised less contentious issues at the beginning and end of meetings, "thereby avoiding parties walking away with a bad taste in their mouth."[81]

The talks gained momentum as trust between the two sides increased. Adherence to a ban on discussing details of the negotiations with the media improved the negotiating climate. The complexity of the issues involved meant that any individual concession had to be viewed within the context of the larger package. Revealing details of particular aspects of the negotiations would likely antagonize certain constituencies, undermining efforts to reach a comprehensive pact. State officials also varied the location of meetings and encouraged casual dress to reduce the formality of negotiating sessions. Michael Finnegan, who orchestrated the talks in his role as governor's counsel, observed that creating a congenial negotiating environment allowed the parties to overcome their initial mistrust: "By insisting on adherence to these simple rules, tension was reduced, the beginnings of trust were established, and substance began to triumph over rhetoric."[82]

The intensive nature of the talks also helped bring the two sides together. Ruzow and Gelber both recalled an incident early in the negotiating process in the summer of 1995 when the two sides lunched at an outdoor restaurant, but at distant tables. Commissioner Gelber ordered a pitcher of beer sent to the watershed table. Initially, the Catskill contingent questioned whether they should accept something from New York City, but their human instincts won out, and they ordered a round sent to Gelber's table. The two sides soon took to sitting together at lunch and developed a rapport that carried over to the formal negotiations.[83] Anthony Bucca offered a telling comparison: "And so it's nice, and we maintain a great deal of cordiality with the commissioner and the people who come from the city. It's actually like the Patty Hearst syndrome: We're greatly enamored of our captors, and they are, personally, for the most part, nice people."[84] Bucca and his watershed colleagues may have felt like hostages during the seemingly interminable negotiating sessions that continued well into fall, past the governor's four-month deadline, but they were hostages with considerable leverage. Although the coalition suspended the lawsuits it had filed, it reserved the right to renew them if negotiations failed.

At the urging of the state government, both sides agreed to let environmentalists join them at the table in July 1995, even though they viewed activists with mistrust. Riverkeeper, a Hudson River Valley–based organization led by Robert F. Kennedy Jr., had sued the city several times for its failure to enforce watershed regulations.[85] Watershed interests also distrusted Kennedy because of his reflexive pro-regulatory bent. In their view, he was a "Fifth-Avenue environmentalist" who had little regard for the economic viability of the Catskills.[86] Nevertheless, both sides recognized the need to include the environmental community to reduce the likelihood of legal challenges to a settlement. By the time Kennedy and other environmentalists joined the talks, the two sides had agreed on many key issues. Although eager to include the environmentalists, the state had no intention of letting their participation derail the negotiations. Michael Finnegan brought them into the discussion on the condition that none of the agreements reached to date, which mostly concerned the land acquisition process and a new regulatory framework, would be reopened for discussion.[87]

These conditions reflected the emergence of a new mode of environmental governance premised on negotiations and the recognition that abstract standards established by unaccountable bureaucrats could not resolve real-life conflicts over resources. The Progressive Era vision of scientific management in which expertise left no room for community priorities and values was now widely recognized as unworkable. Accustomed to fighting for enforcement of existing laws in the courts, some conservation organizations had a difficult time adjusting to the new political terrain, which required much more political agility on their part and a willingness to engage with those whom they had long deemed their enemies.[88]

In the end, reaching a provisional settlement required all parties involved to rethink some of their core assumptions. When he joined the talks, Kennedy frequently compared the Catskills to the highly developed Croton watershed, for which New York had agreed to construct a filtration plant. Fearful that increasing development in the Catskills would inevitably degrade water quality, he initially opposed an economic development package to compensate the region for agreeing to extensive land acquisition by the city and tighter regulations. However, Kennedy began to soften his stance after the coalition convinced him that the mountainous terrain west of the Hudson and the relatively sparse population base bore little resemblance to the suburban sprawl that prevailed in much of the Croton watershed. From the perspective of Catskill residents, many of the gravest threats to water quality had actually decreased in recent decades due to the disappearance of factories and the fading of the dairy industry. Coalition negotiator Alan Rosa recalled how the area had changed since his boyhood days: "I can remember certain times when the Bragg Hollow Stream would just run red from the cows being in the stream. That doesn't happen today. That's a pretty clear stream."[89] In early October 1995, Kennedy made a crucial concession when he endorsed the idea that New York should provide economic development funds for the Catskills.[90]

The city also came to see watershed residents in a new light. Rather than viewing them as potential polluters, it began to see them as partners in water protection. Instead of depending on mechanical filtration, Commissioner Gelber opted "to rely on human beings and relationships between communities to protect the system."[91] Catskill residents took the greatest risks, agreeing to drop their lawsuits against the city in exchange for a settlement.

The provisional agreement announced in November 1995 launched the nation's most aggressive watershed protection program and transformed relations between the city and watershed residents. Lawyers and technical experts would spend the next year hammering out the details of what became known as the Watershed Memorandum of Agreement, or MOA, a thousand-plus-page document. Yet the premise behind the agreement was simple: Catskill residents would partner with New York City to reduce water pollution, but the city would cover the full cost of pollution reduction programs and work to ensure the continued economic viability of the Catskills. In all, New York committed to spend more than $1.5 billion to enhance water quality in all its watersheds, an unprecedented sum for watershed protection, but billions less than the anticipated cost of filtering water from the Catskill and Delaware systems.[92]

The keystone of the MOA was a compromise over the city's right to acquire additional acreage in the Catskills. From New York's perspective, purchasing land represented the most effective way to protect water quality, because undeveloped land would filter out nutrients and pollutants that threatened to degrade Catskill streams. But the era of eminent domain was over—watershed residents insisted on the right to determine the future of their property. Under the so-called "willing seller / willing buyer" provision, the city could purchase watershed property only from landowners who agreed to sell, and it had to pay market rates for any property it acquired. This arrangement ensured that locals would not lose their homes against their will, as occurred during earlier periods of waterworks construction.

However, the flexibility of willing seller / willing buyer also threatened to undermine the city's efforts to improve water quality through land acquisition. The pact proposed tripling city-owned acreage in the watersheds to reduce development pressure, but effectively left the details of land acquisition to the whims of the private market. To ensure substantial land acquisition, the EPA required New York City to solicit roughly a third of the acreage in the Catskill/Delaware watershed, prioritizing the most hydrologically sensitive areas. For all its sophisticated mapping systems and gobs of cash, the city essentially had to cross its fingers and hope that it could acquire enough strategically located land to maintain exceptional water quality.[93]

Language in the agreement also put pressure on New York to permit hunting, hiking, and other recreational activities on much of the land it acquired. As a result, city land purchases offered the possibility of a rare win-win-win: landowners eager to realize gains on their property could obtain them by selling to the city; undeveloped land would improve the quality of water provided to urban and suburban residents; and Catskill residents would enjoy access to these parcels for recreational purposes.[94] The MOA also gave cities and towns the right to exempt certain areas, generally densely developed village centers, from city purchase. As Michael Finnegan observed, the MOA's land purchase provisions represented "a seismic shift in watershed relations that gives form to the City's recognition of the resentment generated by the use of eminent domain."[95] Almost a century after New York City began to lay claim to wide swaths of the mountains, sometimes at bargain prices, the rise of a new political ecology rebalanced the property regime in the Catskills.

The MOA ensured that the Catskills would remain a living watershed. Minimizing the impact of this population on water quality was a core component of the agreement. The coalition agreed to accept a revised set of water quality regulations that limited development in portions of the watershed. The regulations required that all watershed sewage treatment plants install the most advanced technology available within five years, but stipulated that New York City would pay for such improvements. Updating wastewater plants was an expensive but proven means of improving stream quality, the low-hanging fruit of the agreement. The most controversial regulations limited the placement of new septic systems, roads, and parking lots near watercourses and wetlands, and required city approval of a storm-water pollution prevention plan for most new commercial and industrial projects.[96]

In exchange for agreeing to the more stringent watershed regulations, the coalition extracted a substantial package of concessions—formally known as Watershed Protection and Partnership Programs. New York City agreed to provide $270 million to watershed communities to help them comply with the cost of the new regulations, and to compensate them for diminishing the region's development potential. The funds would cover the cost of new sewage treatment plants, improved sand and salt storage to reduce deposition into streams, septic system upgrades and replacements, and a host of other water quality enhancements. Most of these programs would be administered by a new agency, the Catskill Watershed Corporation (CWC), an organization based in the watershed. Staffed and overseen primarily by Catskill residents (the New York City mayor and the governor each selected a board member, but the staff and remainder of its board hailed from the Catskills), the CWC put a local face on the agreement. In addition to administering programs such as septic system replacements and storm-water retrofits, the CWC would oversee the allocation of the Catskill Fund for the Future, a $60 million loan and grant program intended to promote ecologically sustainable economic development in the Catskills. A diverse group of watershed business owners was eligible for loans or grants to expand operations. To ensure the long-term viability of the fund, the MOA required that most of the money be distributed in the form of loans.[97] The partnership programs also included almost $10 million in "Good Neighbor" payments. Financial gravy that eased the way for a deal, the payments amounted to free money for Catskill towns to invest in capital projects of their choosing. The link between Good Neighbor payments and improved water quality was purely political. The CWC argued that the subsidies would "help establish a better working partnership with communities in the Watershed."[98]

Money alone was not enough to win the coalition's backing for a deal. While watershed officials craved the infusion of funds the agreement would provide to their towns, they questioned the city's willingness to embrace a truly cooperative approach to watershed management. Creating an institutional framework that would allow Catskill officials and residents to manage a broad array of watershed programs and provide them an opportunity to challenge city decisions emerged as a critical component of the MOA.

The most important new entity created by the MOA, the Catskill Watershed Corporation, was modeled on the highly successful watershed agricultural program. Headquartered in the Catskill village of Margaretville, the CWC was the organizational heart and soul of the watershed program. Establishing a locally based agency to work with Catskill residents on water quality projects represented a risk on the part of New York City, but, as Commissioner Gelber observed, the success of watershed protection required the active cooperation of watershed residents: "Watershed communities have to be your partner in protecting the system, and there's no amount of mechanical systems or science that's going to protect the water unless people in the watershed are your partner in protecting it."[99]

Well aware of the long history of tension between the watershed and the city, MOA negotiators created a new organization, the Watershed Protection and Partnership Council (WPPC), to serve as a forum for settling the disputes that would inevitably arise. The brainchild of Governor's Counsel Michael Finnegan, the WPPC brought together representatives from community groups, the city, state, and watershed communities from both sides of the Hudson to discuss the progress of watershed protection. Intended to forestall the litigation and political battles of the early 1990s, the council provided a public venue at which the interested parties could hash out disputes before they escalated. As coalition attorney Dan Ruzow noted, the WPPC meetings also served as a way for watershed communities to hold New York accountable for its actions: "In the past, the city didn't have to talk to anybody. There was no place where you could force the city to come publicly and talk, to say why they were doing something. But if you have a partnership meeting and the city's out doing something, you can bring to the table, 'City. Why are you doing this?' So it empowers those groups that have less power. Again, it's a very useful thing."[100] After spending many months hashing out an agreement, both sides were well aware of the need for an orderly means of resolving disputes.

The agreement signaled the beginning of a new relationship that struck a balance between urban needs and rural preservation. Optimism was the order of the day on November 2, 1995, when environmentalists, coalition members, and government officials gathered at the World Trade Center in lower Manhattan to announce a preliminary deal. The *New York Times* declared the deal "a rare triumph of good sense over once insurmountable grievances." Its reporter marveled at the newfound comity between longtime enemies, symbolized by the sight of a smiling Robert Kennedy Jr. ("who for many years had been openly reviled by many watershed residents") standing alongside coalition activist Anthony Bucca.[101]

A true compromise, the MOA did not fully satisfy any of the parties, yet managed to meet the magic threshold of acceptability. The state saved New York City from the expense of filtration while at the same time protecting the economic and social viability of the watershed. The terms of the agreement also passed muster with federal regulators. The combination of land acquisition, stricter regulations, and a suite of new infrastructure projects seemed likely to improve water quality. If the city's water failed to meet federal standards, the EPA reserved the right to rescind its waiver and mandate construction of a filtration plant.

The big winners, of course, were the city and watershed residents. Mayor Giuliani boasted that the settlement would increase water rates by only seven dollars a year for the typical family, instead of the hefty rate hikes needed to finance a filtration plant. For its part, the coalition could point to the willing seller / willing buyer provision, the substantial economic development package, local control over most of the new watershed programs, and the city's acceptance of its responsibility to fully finance the cost of watershed protection as proof that Catskill residents had finally forced New York to meet its social and financial obligations to the watershed. Coalition executive director Ken Markert gauged the success of the pact by comparing it with Commissioner Gelber's initial settlement offer in the fall of 1994: "Now, with the active support of the state, we've got this deal that's worth at least ten times that."[102]

Same Place, New Visions

The signing of the formal Watershed Memorandum of Agreement in January 1997 capped a remarkably tumultuous period in the city's water supply history. Resistance from an array of stakeholders—federal and state governments, environmentalists, and watershed residents and officials—forced New York to fundamentally alter the way it conceptualized and operated its water system. The city had traditionally taken its own political power and the abundance of nature for granted. By the mid-1990s, many municipal officials recognized the limits of natural systems and the need to nurture relationships with traditional antagonists. The expansion of environmental regulations at the state and federal level was largely responsible for this radical shift in perspective. Watershed residents, their attorneys, and visionary city officials played critical roles in crafting the details of the MOA, but they would have remained backstage if federal regulations had not required New York to filter its water supply. Similarly, the possibility that the state would prohibit additional sewage hookups in parts of New York City substantially increased the DEP's enthusiasm for toilet replacement and water conservation.[103]

Traditional regulatory controls may have provided the backdrop for the watershed negotiations, but it was government's speaking role—its power to appease, broker, and persuade—that ultimately led to an agreement. The MOA negotiations reveal an important shift in environmental governance that crystallized in the 1990s. The mountain of environmental regulations enacted in the 1970s reduced point sources of pollution, leading to cleaner air and water. But "command-and-control regulations" proved a blunt instrument; by the 1980s their limitations had become clear. Not only did they largely ignore the increasing role of nonpoint pollution in fouling the nation's air and waters; they frequently generated endless rounds of litigation, thwarting efforts to devise practical solutions to complex environmental problems. State and federal governments began to rely on a combination of traditional regulations and negotiations in an effort to avoid protracted legal battles. In western states, the federal government mediated the conflict between private property rights and endangered species, devising innovative habitat conservation plans. The MOA revealed a similar flexibility on the part of government. In this case, the federal government supplied the traditional top-down regulations, and the state government oversaw the nitty-gritty negotiating.[104]

The MOA negotiations also highlighted the contingent power of government, an important theme in recent studies of American politics. Federal law clearly gave the EPA the power to mandate construction of a filtration plant for Catskill water. However, this heavy-handed use of government power would have enraged authorities in New York City and led to a series of legal battles. Devising the source-protection alternative required a much defter application of government authority. Commissioner Gelber's accessibility and the election of George Pataki as governor fostered a positive environment for negotiations. The MOA is a welcome reminder that government is less an abstraction than a changing cast of characters with particular biases and priorities.[105]

The embrace of a collaborative approach to environmental management forced municipal authorities to radically revise their conception of the water system. For decades, the Board of Water Supply, the Departments of Water Supply and Environmental Protection, and municipal leaders viewed the water system either very close up—at the level of pipe and valve—or from a great distance, like an Impressionist painting that only came into focus from afar. The shift to source protection and conservation forced managers to attend to new details and to master the middle ground they had previously overlooked. Plumbers replaced toilets in hundreds of thousands of apartments. DEP employees created detailed maps of thousands of watershed parcels to guide their land acquisition decisions. They then had to forge relationships with thousands of landowners in an effort to persuade them to sell their property to New York City. A successful source-protection program would require New York to develop an intimate familiarity with the million-plus acres of the Catskill/Delaware watershed, not just the sliver of it under city ownership. The watershed, which in the eyes of generations of officials had been a coherent, largely undifferentiated mass, became a jigsaw puzzle of streams, villages, farms, and property owners.

The shift to a conservationist mind-set also implied a new relationship with technology. New York's long love affair with upland water sources produced a complex series of dams, reservoirs, and aqueducts. A by-product of this devotion to system expansion was intense resistance to technologies such as water metering and filtration. The MOA and the embrace of water conservation signaled a sharp departure from the rigid insistence on massive reservoirs and elegant dams. High technology still had its place; sewage treatment upgrades and construction of the Croton filtration plant were critical components of New York's efforts to protect and improve water quality. What changed was the city's embrace of low-technology solutions. Replacing toilets and installing meters did not directly reshape the landscape, but water conservation saved hundreds of millions of gallons a day, as much water as a new reservoir would provide. Similarly, while land acquisition involved the use of sophisticated computer programs to prioritize parcels, it was based on the idea that nature, not a human-made machine, was the most effective water filter available.

The radical reorientation of New York City's water policies also reflected changes within municipal government. The elimination of the BWS in 1978 silenced the most vigorous supporter of water system expansion. The DEP was slow to chart a new course. However, by the early 1990s, as Bureau of Conservation Services director Warren Liebold observed, the agency had changed substantially, transforming itself from a traditional government agency in which politics trumped efficiency to a well-managed "business organization that has to collect and raise money, deliver services, and accomplish particular goals."[106] The appointment of Albert Appleton, the first non-engineer to lead the DEP, ensured that the emphasis on rational management and receptiveness to new ideas extended to the upper echelon of the agency.

Ultimately, it was unyielding pressure from the state and federal governments that enabled Marilyn Gelber and water conservation activists at the DEP to prevail. The combined forces of politics, limited finances, and ecology severely constrained the city's options. In the 1990s, state government refused to tolerate New York City's attempt to unilaterally impose its will on the Catskills. Building on its support of new reservoir release regulations in the 1970, the state insisted that a new deal in the watershed must reflect the interests of those who lived there. Environmental regulations also became much more stringent and comprehensive in the intervening decades.

In the new era of political ecology, concerns that were once deemed marginal gained prominence. Scientific advances convinced Congress that the threat posed by unfiltered water justified the enormous expense of filtration. Political norms also changed radically. The objections of Catskill residents, which state officials had almost entirely disregarded in the 1950s, emerged as the greatest obstacle to a vigorous watershed protection program. The MOA ended the long reign of urban autonomy in the mountains.

The great paradox of watershed protection, of letting nature do the work, was that it required an intensive engagement with a diverse group of stakeholders on a wide range of issues. At the same time that the MOA provided mountain residents a direct voice and role in managing the watershed, it also greatly increased the presence and power of New York City in the Catskills. An emphasis on source protection entailed a heightened environmental police presence by the city, additional monitoring of watercourses by DEP employees, DEP oversight of many routine construction projects, increased city land ownership, and a host of other intrusions. Nevertheless, active state and federal oversight, continued activism by watershed residents, and disclosure and consultation requirements mandated by the MOA circumscribed the city's authority.

The $1.5 billion question was whether these measures would improve water quality. Hammering out a deal had been extraordinarily difficult. Implementing it would prove equally challenging.

CHAPTER SEVEN

Implementing the Watershed Agreement

In the spring of 1997, a few months after signing the Memorandum of Agreement, New York finally received the prize it had desperately sought: a five-year EPA waiver exempting the city from filtering its mountain water supply.[1] Its immediate future seemingly secure, the water system no longer made front-page news. The reality was much more complex. In order to renew its waiver, New York would have to initiate a broad-based watershed protection program that required it to solicit hundreds of thousands of acres of privately owned land for purchase, upgrade sewage treatment plants, and implement new regulations. The watershed pact did not liberate the DEP from federal regulations; instead, it put the agency on lifetime probation.

Protecting water at its source posed three main challenges for the agency. The first order of business was to greatly expand and diversify its workforce. Prior to the 1990s, few employees were based in the watershed. A source protection approach required the DEP to establish a presence throughout the Catskills and hire people with titles not typically associated with water departments: forester, real estate manager, and geographic information systems specialist. Assembling this team, clearly defining their tasks, and getting them to work together was a tall order. As significant as these internal changes were, the DEP could not improve water quality without the assistance of its partners in the Catskills. The MOA required the city to fund nonprofit organizations whose mission was to reduce water pollution while maintaining a vibrant local economy. The Watershed Agricultural Council (WAC) continued to work with Catskill farmers to minimize the flow of agricultural pollutants into streams, and initiated a program to permanently preserve farmland through conservation easements. Motivated by a building moratorium, Delaware County officials launched an ambitious plan to reduce pollution entering the Cannonsville Reservoir. The Catskill Watershed Corporation was the centerpiece of these local efforts. The CWC distributed grants and loans to area businesses, managed septic and storm-water programs, and worked to cultivate positive relations between watershed interests and New York City.

The third challenge—land acquisition—was the biggest wild card. The DEP hired a team of experts to prepare maps that prioritized the most hydrologically sensitive properties. Inevitably, owners of desirable parcels often proved unwilling to sell to the city, forcing New York to settle for less critical land. The mismatch between the orderly urban vision of a protected watershed and the messy reality of the willing seller / willing buyer provision underscored the dramatic shift in power relations unleashed by the MOA.[2]

The new balance of power reconfigured both rural and urban landscapes. The erection of roadway signs bearing witness to communities displaced by reservoir construction signified an end to the erasure of watershed history in the Catskills. The descendants of those flooded out by New York City were, for the first time, permitted to hunt and sail on DEP property. Dairy farmers moved their cows from the barns to the pasture, embracing a more traditional form of agriculture that promised to increase profits and improve water quality. The need for clean water also reshaped the urban built environment. A not-in-my-backyard dispute over the proposed location of the Croton Filtration Plant yielded an unlikely outcome: substantial renovation and rehabilitation of parks throughout the Bronx. The filtration plant promised safer water for residents connected to the Croton system. Sewage treatment upgrades and improved land management practices also ensured higher-quality supplies for the vast majority of New Yorkers whose water originated in the Catskills.

Upbeat press releases trumpeting city-funded economic development in the Catskills and increased recreational access to newly acquired watershed lands obscured the fragility of the urban-rural relationship. Despite significant changes in personnel and a strong commitment to watershed protection, New York did not immediately shed the habits that had long angered mountain residents. It continued to challenge tax assessments of its watershed holdings, depleting a $3 million fund established to help watershed towns cover the legal costs of tax disputes. Catskill residents also worried that city acquisition of large swaths of the watershed would reduce public access to these lands and stifle economic growth. Renewed tensions between watershed residents and New York City threatened to stunt the most ambitious watershed protection program in the United States before it could bear fruit. Unable to resolve disputes related to taxes, land acquisition, and other issues, watershed residents once again turned to the courts to rein in the city.

A major external threat also loomed. New technologies whetted the appetite of energy companies for the natural gas locked in shale deposits throughout the Catskill-Delaware watershed. City officials feared that large-scale extraction of natural gas could contaminate the water supply they had worked so hard to protect. The future of New York's water supply, seemingly secured with the signing of the MOA in 1997, now appeared very much in doubt.

Signs of Change

Many of the changes in the Catskills were invisible to the casual observer. A worker installed a new septic system on a remote property; a landowner agreed to sell her woodlots to New York City; a struggling farmer reduced the amount of grain he fed his cattle in favor of more pasturing—the cumulative effects of these individual actions would determine the fate of watershed protection. In other respects, the impact of the MOA was clearly imprinted on the natural and built environment of the Catskills. One way to gauge the cultural, psychological, and physical impact of the agreement is to examine the changes along Catskill roadways from 1999 to 2005.

In the spring of 1999, the DEP, in conjunction with the state Department of Transportation, installed road signs at the borders of the Croton and Catskill-Delaware watersheds. A typical sign posted near the Catskill Mountain town of Stamford, New York, read:

Water Supply Area
Next 27 Miles
Report Polluters
1–800–337–6921

From New York's perspective, the signs delivered a clear and uncontroversial message: Help us protect the watershed. The signs used standard wording, varying only in the mileage listed, which indicated the extent of the watershed the driver could cover on the road posted. Controversy erupted primarily in Delaware County, in the agricultural heart of the Catskills.[3]

Town officials and residents strongly condemned the latest addition to county roads. Jim Eisel, a supervisor in the town of Harpersfield and a CWC board member, claimed that the signs reminded him "of communist Russia. The city is asking neighbors to turn in neighbors." The president of the county's chamber of commerce, Ray Pucci, echoed Eisel, observing, "It's like we're in a police state."[4] Even the CWC, the city's most prominent partner in the watershed, cried foul. As its newsletter observed, residents objected to the signs because of their "harsh language" and because the DEP failed to consult local officials before erecting them.[5] The lack of consultation particularly irritated officials such as Ray

Christensen, chairman of Delaware County's Board of Supervisors. The county was planning to allocate half a million dollars for stream stabilization to reduce sediment loading in watercourses that emptied into the Cannonsville Reservoir; the least the city could do was act like a good neighbor.[6]

The DEP's response to the signage flap revealed an agency struggling to adapt to a revamped political landscape. Joel Miele, an engineer who succeeded Marilyn Gelber as DEP commissioner, dubbed the dispute a "tempest in a teapot."[7] In the pre-MOA era, the city likely would have resisted changing the signs, leaving them vulnerable to vandalism. However, in the post-MOA era, the DEP proved more nimble. The unanimous condemnation of the signs by local officials persuaded the agency to create a multi-jurisdictional task force to devise a new design and wording. City officials wisely concluded that cooperating with local officials was the most effective means of safeguarding water quality. The new signs were smaller than the original ones and cast watershed protection in a much more positive light. They read, "Welcome to Our Watershed: Help Keep It Clean." Instead of urging motorists to snitch on polluters, it urged them to report "problems" to 1–888-H2O-SHED, a number drivers were more likely to remember.[8]

The pollution signs were part of a larger effort to sensitize residents and visitors to their presence in the watershed. Catskill residents did not need to be reminded of the reservoirs and the city's outsize role in daily life, but visiting motorists could easily zip through the region, unaware that massive reservoirs lurked behind a thin curtain of trees. Many tourists only learned of the urban presence when they got out of their cars and spotted the "City of New York: No Trespassing" signs posted on reservoir buffer lands. By 2004, more plaintive signposts began to dot the Catskills. That summer, the CWC erected twenty-six road signs marking the former locations of towns submerged by reservoir construction. These modest brown and yellow signs acquired a powerful resonance through accumulation; even a relatively inattentive motorist could perceive the widespread geographic and social impact of reservoir construction upon spotting the fifth or sixth reminder of a vanished community. In 2002, the CWC built sturdy informational kiosks at each Catskill reservoir detailing the history of vanished communities and the development and operation of the city's water system. The dedication ceremonies for these kiosks attracted former community residents and retired Board of Water Supply personnel.[9]

These low-cost steps to memorialize the region's pre-watershed past marked a decisive shift in the city's portrayal of watershed history. For decades, the board trumpeted the benefits of system expansion and modernization. By focusing on both the history of submerged villages and the engineering of the water system, the CWC suggested that the past was more complex. Almost a century after former mayor George McClellan rhapsodized about the benevolent modernizing of the countryside, the other side of the story finally gained official sanction.

This rebalancing of history suggested that a more nuanced conception of modernity had taken hold in the watershed. Social theorist Marshall Berman argued that a central feature of the modern mind-set was its denigration and erasure of the past.[10] The board's mission to continuously expand the water system perfectly embodied this philosophy: newer was better. In the post-MOA era, the past acquired new standing. Acknowledging previous hardships associated with waterworks construction had the potential to promote more positive working relations between DEP and watershed residents. Most important, the fundamental premise behind the agreement—that watershed protection and economic vitality could coexist—required the careful blending of stability and change. The MOA empowered New York to purchase substantial acreage in the watershed to prevent forest fragmentation and storm-water pollution linked to expansion of impervious surfaces. But it also required the city to invest tens of millions of dollars to promote business development in the Catskills. For decades, New York focused on building advanced dams and aqueducts, assuming that the Catskills would somehow remain untouched by the modern world. The MOA ended the practice of relying almost exclusively on the geographical isolation of the Catskills to ensure the purity of the water supply. The price of maintaining and expanding an "old-fashioned watershed" dominated by fields, farms, and forests was extensive and expensive modernization of Catskill homes, businesses, and communities.

One of the major outgrowths of the economic development projects funded by the MOA was the sprucing up of Catskill hamlets and villages. Businesses and communities tapped CWC grants and loans to upgrade aging infrastructure. One round of funding awarded in 2004 financed more than twenty separate projects, including repairs at a local museum, facade improvements for three commercial buildings, large bluestone signs at the entrance to the village of Tannersville, and the purchase of stage equipment and safety rails at a large theater.[11]

In addition to beautifying outdated infrastructure and enriching the region's cultural offerings, watershed economic development funds helped many area residents establish new businesses. South Kortright residents Daniel and Diane Dax received a low-interest loan to renovate a century-old carriage house that quickly became one of the most popular wedding and facility destinations in the region. In this instance, as one appreciative resident observed, modernization and preservation went hand in hand: "I'm not sure sainthood would be a sufficient reward for saving that beautiful building. Most people would have just called in the bulldozers. The main supports were rotted through and the posts had fallen through the floor to the ground. It was a bold choice with a beautiful result that shows what can be accomplished."[12] Forty-five miles southwest, in the town of Lanesville, Amy Jackson opened a new restaurant with support from the CWC. Jackson's chosen location, a former general store and post office near a small stream whose waters eventually empty into the Ashokan Reservoir, proved challenging. Because watershed regulations generally prohibited construction within one hundred feet of a watercourse, she spent four years applying for variances and scraping together funding to construct the required septic and storm-water systems. Despite the DEP's refusal to cover the full cost of the storm-water system, Jackson managed to open the restaurant in 2006. CWC funds also enabled her to upgrade the building facade and erect an attractive sign. Jackson's experience illustrated the paradox facing Catskill business owners in the post-MOA era. Watershed regulations constrained businesses' ability to expand or relocate, but CWC economic development funds also offered the possibility of economic growth. Several years after its establishment, Amy's Take-Away was flourishing.[13]

Although individual businesses benefited from the grants and loans disbursed by the CWC, even the most optimistic observers acknowledged that the watershed agreement did not fundamentally transform the Catskill economy. The features that make the region an attractive water source—its remoteness, rugged topography, and low population density—also discourage economic investment. This is particularly true of Delaware County, where the Cannonsville and Pepacton Reservoirs are located. Watershed Agricultural Council chair Fred Huneke offered a clear-eyed assessment of the county's economic base: "We don't have any major highways, we don't have an airport, we don't have any cities . . . so what do you have left? You have agriculture, you have forestry, natural resource-based industries, you have tourism, and that's just about it."[14]

The economic challenges in Delaware County resembled those in other upstate areas. Even those who have made a career out of challenging the city's prerogatives acknowledged that the reasons for the region's sluggish economic growth were largely unrelated to watershed regulations. Taking stock of the MOA in 2011, longtime Coalition of Watershed Towns attorney Jeff Baker noted that watershed communities have generally outperformed their rural counterparts in other parts of New York's southern tier:

> There's a strong argument that being in the watershed in a lot of ways helps things because you do have this money that's coming to communities to build wastewater treatment plants. No one's getting grants like that. You've got money for the septic systems. It's not just helping the contractors, which of course it is, but it's helping landowners. People are getting either free or heavily subsidized septic systems. You have more of a regulatory burden; there's no doubt about that. And that does certainly create a burden. But there's some compensation for that, some of those costs are covered.[15]

The MOA enhanced the competitiveness of agriculture and provided substantial employment opportunities for contractors throughout the watershed, but it did not markedly increase tourism in the region. Individual counties maintained separate tourism promotion efforts, diluting the regional identity of the Catskills. Second-home owners seeking a refuge after the September 11 terrorist attacks flocked to the Catskills and other rural areas within easy driving distance of New York City. Although home purchases and renovations stimulated watershed economies, they did not raise the low profile of more remote portions of the watershed. In a region whose economy once revolved around the boarding business, CWC executive director Alan Rosa could only lament that in much of the watershed, "There is no place to stay."[16]

Recreation in Country and City

If the economic effects of the MOA were ambiguous, the recreational repercussions were not. One of the most contentious issues in the watershed negotiations concerned recreational access to city-owned land. The DEP tightly regulated access to its buffer lands by requiring an array of permits and limiting the types of sanctioned activities. Prior to the MOA, most city watershed property consisted of land surrounding its reservoirs. Restricting access to these modest holdings (the city owned approximately 35,500 acres of buffer lands before 1997, only 3.5 percent of the Catskill-Delaware watershed) did not represent a hardship for watershed residents, who could hike and hunt on private lands and on state-owned property in the Catskill Park.[17]

The scale of land acquisition envisioned under the watershed agreement portended significant recreational changes. If the city purchased tens of thousands of acres, as seemed likely, residents could potentially lose access to many prime recreational parcels. Although many residents posted their property, they often permitted relatives and friends to bird watch, hunt, hike, and fish on their lands. Anticipating resistance from the DEP, coalition negotiators inserted language in the MOA that required New York to make lands it acquired available for historical recreational uses.[18]

The vagueness of this language and the DEP's wariness of people tromping across its newly acquired watershed properties ensured that recreational access remained a source of tension. In a paradox of political ecology, the managerial path of least resistance—declaring most city-owned lands off-limits to recreation—represented the most politically fraught approach. Watershed residents who had expected to reap recreational benefits from these purchases soon learned that the DEP was generally unwilling to open up these newly acquired lands. The September 11 terrorist attacks only reinforced the agency's concerns about contamination of its water supply. In addition to fretting about careless hunters and hikers, the DEP now worried that its reservoirs were an appealing sabotage target.

Despite the legitimacy of concerns about deliberate contamination of the water supply, the agency's insistence that the possibility of terrorism required limiting recreational access to watershed holdings made little sense. The decision to temporarily ban all public access to its watershed lands and reservoirs after the attacks unnerved Catskill residents, who worried that the policy would fundamentally change their relationship with nature. Holly Aplin, who lived just south of the Schoharie Reservoir, accused New York of targeting the wrong people in its zeal to safeguard water quality:

> I am struck with a sadness as to what the Catskill Mountains are becoming. Will my grandchildren be able to play in fields, swim and fish in the streams, or hike and explore the mountains as I did as a child? Probably not, without fear of being arrested. People living in the area designated as "the watershed" need to open their eyes to the control New York City has placed over this beautiful area. . . . Because of the 9/11 disaster, we won't be able to fish in the reservoir. Does NYC really think a fisherman is going to harm the water?[19]

The DEP lifted its recreation ban in 2002, but maintained its onerous permit system, arguing that after 9/11 it needed to know who was using its lands. CWC director Alan Rosa and others recognized that the patchwork nature of city holdings made it impossible to effectively monitor activity across the sprawling watershed. Rosa argued that a permit system would not deter terrorist activity. As he dryly observed, "If a terrorist is going to do something, he's not going to say, 'Oh, I forgot my DEP permit,' and go get a DEP permit and come back."[20]

The agency struggled to reconcile its concerns about water contamination with the potential political benefits of expanding recreational access. A 1999 decision to provide access to seven newly acquired properties revealed these tensions. The city did not plan to maintain trails on any of the new lands, effectively limiting access to only the most intrepid explorers. Those who wanted to both hike and fish on the new properties had to obtain two separate permits. Commissioner Miele's statement announcing the decision reflected the agency's ambivalence about opening up its holdings: "It is important that visitors to city properties observe these rules to demonstrate that recreational uses on city-owned lands can be compatible with protection of water quality in the streams and reservoirs of the watershed."[21]

In at least one instance, the DEP's interests dovetailed with those seeking greater access to watershed lands. The agency relied on a heavily forested landscape to naturally filter its drinking water, and planted trees on many of the parcels it acquired. The tendency of deer to browse young trees thwarted reforestation. From the city's perspective, encouraging deer hunting thus made eminent sense. By thinning the deer population, hunters could help protect water quality. In 2002, the DEP announced that it would permit deer hunting on almost twenty-four thousand acres, nearly half of which had not previously been open to hunters.[22] Expansion of deer hunting proved extremely popular with Catskill sportsmen. CWC director Rosa recalled a conversation he overheard in the town of Margaretville in 2010. "I was at the Hess station up here getting gas and there was two guys out there talking about hunting. One guy was successful and got a nice buck on city land. He said, 'Opening up that city land . . . this is the greatest thing that ever happened.'"[23]

In the eyes of many watershed residents, the expansion of deer hunting was the exception that proved the rule. They grew increasingly frustrated with the seemingly arbitrary nature of the agency's recreation decisions. Hunters questioned the DEP's willingness to permit deer hunting on many of its holdings while continuing to ban the taking of other types of small game.[24] State officials also criticized the inconsistency of these policies. Why, they wondered, did the city prohibit recreational activities on its own watershed lands that the state permitted on the more than two hundred thousand acres it owns in the New York City watershed?[25]

Until 2006, the DEP resisted significantly expanding recreation on its growing portfolio of watershed holdings. But Emily Lloyd, who was appointed DEP commissioner in 2005, proved much more willing than her predecessors to increase recreational opportunities on city-owned lands. In the summer of 2006 Lloyd made land-use permits instantly available through the DEP website, opened some of its holdings to the general public (no permit is required to access these lands), and legalized the hunting of small game, turkey, and bear on thousands of acres. The agency also further expanded the acreage open to deer hunting to include almost half of its publicly accessible lands. Local officials heralded the benefits of instant permit application for the Catskill economy. The change had "been a long time coming," observed one town supervisor, who held out hope that "this will be a great economic tool for local businesses, and . . . benefit residents, sportsmen, hunters and visitors."[26]

The loosening of restrictions in 2006 marked a decisive turning point. Since 2006, the DEP has continually expanded recreational access to its watershed holdings. In 2008, the agency finally succumbed to pressure from state officials and opened up almost eleven thousand acres of city watershed property adjacent to state-owned land to hikers, hunters, and fishermen without need for DEP permits, making these parcels functionally equivalent to state conservation land. In 2009, it initiated a pilot program to permit recreational boating of nonmotorized craft on the Cannonsville Reservoir. The agency anticipates expanding the program to other reservoirs. By 2011, visitors could hunt and hike almost two-thirds of the DEP's watershed lands without applying for a city permit.[27]

The DEP proved more willing to compromise on recreation than on any other aspect of watershed management. Alan Rosa argued that the change of heart was a product of simple experience: "As time went on, and they seen that, you know, there wasn't a bunch of bogeymen out there that was going to ruin the land, they kept opening up more land."[28] On this issue, at least, the apparent conflict between city and country proved illusory; both sides benefited by expanding recreational access.

The water system's impact on recreation in the new century extended from the Catskills all the way to the Bronx. Although the DEP also acquired land in the Croton watershed to minimize source pollution, its primary strategy for improving water quality in the region was to build a filtration plant. Intense community opposition derailed the agency's plan to build the plant on the site of the Jerome Park Reservoir in the Bronx neighborhood of Williamsbridge. The DEP then settled on Van Cortlandt Park as its preferred location. The revised plan called for building the plant under the Moshulu Golf Course, in the southeast quadrant of the park. This idea also encountered resistance, both from neighborhood residents who opposed the traffic disruption from a daily parade of construction vehicles and the presence of industrial chemicals in their midst, and from park advocates, who decried the loss of open space.[29] A decade after signing the consent decree with the federal government, the city had made no progress toward filtering the Croton supply. Determined to move the project forward, in 2003 Mayor Michael Bloomberg made a bold offer to secure the support of elected officials in the Bronx: in exchange for permission to build the plant in Van Cortlandt Park, New York would invest $200 million over five years to rehabilitate parks throughout the Bronx, three times the expected appropriation level. The generous funding persuaded state legislators to support his plan. The following year, the city signed an agreement with the state committing to these recreation investments.[30]

Parks in virtually every Bronx neighborhood benefited from the mayor's gambit. The improvements fell into several categories: upgrading neighborhood parks, renovating regional recreational facilities, improving the waterfront, and greening the borough through projects such as street tree planting and expanding plantings in parks. Parks Commissioner Adrian Benepe called the rehabilitation program "a renaissance in the Bronx unparalleled since W.P.A. days."[31] South Bronx parks were among the first beneficiaries of the program. New play equipment, spray showers, basketball courts, seating, and landscaping turned shuttered neighborhood parks into prized community amenities. Before the renovations, Story Playground was a playground in name only: the basketball court lacked baskets, and grass poked through the cracks that riddled the court. Croton mitigation funding utterly transformed Story; its shiny new ball courts and play facilities made it one of New York's most attractive neighborhood playgrounds.[32] In a sign of the water system's long-standing intimate connection to urban parks, Croton funds also financed the overhaul of Aqueduct Lands and the Williamsbridge Oval, two Bronx parks whose origins can be traced directly to the water system. (Aqueduct Lands is a linear park along the route of the Old Croton Aqueduct.) Ironically, development in the suburbs, the driving force behind construction of the Croton filtration plant, bequeathed a green legacy to the Bronx.

These suburban areas also benefited from the DEP's newfound willingness to increase recreational access to its reservoirs and watershed lands. In 2008, the agency opened most of its east-of-Hudson reservoirs to ice fishing. While the DEP's claim that "New York City's reservoirs offer some of the best fishing opportunities in the country" stretched the truth somewhat, the Croton reservoir system does provide excellent warm- and cold-water fishing.[33] Anglers from the metropolitan area who acquired the necessary permits enjoyed reservoirs stocked with black bass, lake trout, and brown trout. No other American city offers millions of citizens such abundant fishing opportunities within an easy drive of their homes. In the 1840s, the ability to transport water forty-one miles from the Croton region

Fig. 15. Story Playground, Bronx, after rehabilitation. Mitigation funding from the construction of a new filtration plant in Van Cortlandt Park paid for the sprucing up of parks throughout the Bronx. (New York City Parks and Recreation Department)

to New York City was considered an engineering marvel. Twenty-first-century New Yorkers were amazed to find that a reasonably short drive from Brooklyn offered a winter tableau that resembled "northern Minnesota or Vermont, maybe even Alaska."[34] Throughout the southeastern portion of the state—from the forests of the western Catskills to the dense network of reservoirs in Putnam and Westchester Counties to the playgrounds and parks of the Bronx—the dual imperative to protect water quality and enhance access to city water supply properties yielded recreational benefits for millions of New York State residents.

Homes, Fields, and Forests

The Croton filtration plant departed from the city's source protection strategy in two principal respects. Its purpose was to cleanse water after it had become polluted, whereas the objective in the Catskills was to prevent the water supply from becoming contaminated in the first place. Second, it was a centralized solution to combating disparate sources of pollution. By opting for watershed protection over filtration in the Catskills, the DEP committed itself to a decentralized pollution-control strategy that relied on a variety of technological and political solutions. The vast expanse of the sixteen-hundred-square-mile Catskill/Delaware watershed (eighty-five miles separate the Ashokan Reservoir in the eastern section of the watershed from Cannonsville at the western edge), funding limitations, and the variety of polluters—households, businesses, farms, roads—required the agency to craft strategies tailored to individual homes, farms, and communities.

This individualized approach to protecting water quality represented a sea change in the city's conception of the Catskills. During the decades when it constructed its mountain water network, the BWS paid scant attention to the topographical, economic, and social complexity of the region. Fixated on a single common denominator—the abundance of water—the board ignored variations in population density and land use. A source protection strategy compelled the DEP to consider the variability of the Catskill landscape in designing and implementing its watershed programs. Upgrading a septic system for a house with rocky soils and a system for a house with sandy soils posed different challenges. Similarly, given differences in topography, stream location, and size of operation, the most effective pollution-control approach for a particular farm might vary substantially from that for a farm two miles down the road.[35]

Cleaning up the waste stream from the thousands of homes and businesses scattered throughout the watershed required a variety of solutions. The first step was to significantly upgrade the five city-owned wastewater treatment plants in the Catskills. Rehabilitating New York's treatment plants made practical and political sense. Because these facilities were the largest in the region—they treated 40 percent of the wastewater in the Catskill-Delaware watershed—modernizing them yielded immediate water quality improvements. Prioritizing city-owned facilities also meant that the DEP was no longer in the uncomfortable position of asking other wastewater treatment plants to comply with standards that its own plants failed to meet.[36]

In 1999, the DEP turned its attention to improving treatment facilities operated by municipalities and private businesses. The MOA mandated tertiary treatment of wastewater in the watershed, necessitating installation of the most sophisticated technology available to reduce phosphorous levels and pathogens. According to the agreement, New York City was obligated to cover the expense of facility upgrades and any operating costs associated with the more stringent requirements. Improvements funded by the MOA built on a wastewater monitoring and assessment program mandated by the EPA as a condition of the first filtration avoidance determination issued in 1993.[37] By the end of 2002, the DEP had upgraded facilities accounting for almost 90 percent of wastewater treatment plant flows in the Catskill watershed.[38]

Although the DEP devoted the lion's share of its budget for waste prevention to upgrading treatment plants, informed observers considered septic repair and replacement a more effective pollution prevention program. The region's low population densities made septic systems the preferred method of waste disposal in the Catskills. Many of these septic systems used outmoded technology or did not function properly, and thus failed to filter out contaminants that eventually made their way into creeks and streams. A study found that approximately 70 percent of Catskill septic systems were located on unsuitable soils. Alan White of the Nature Conservancy observed that "we have a hardpan soil . . . but if you've got a clay layer 18 inches down that prevents percolation[,] septic systems don't work here." The combination of outmoded systems and unfavorable soil conditions made septic replacements critical to continued improvements in water quality.[39]

Replacing several thousand septic systems in the Catskills proved much more challenging than installing hundreds of thousands of new toilets in New York City. At one level, the difference came down to time and money. A plumber could install a new toilet in a couple of hours for under $300; replacing a septic system could take several weeks and usually cost several thousand dollars. Because New York City covered virtually all the expenses of system replacement for full-time residents, the program was extremely popular with both homeowners and contractors. Residents appreciated that taking steps to properly dispose of their sewage would not represent a financial hardship, and contractors enjoyed the reliability of prompt payments from CWC coffers. As the corporation gained experience in administering the program, and the possibility of steady work attracted contractors, the pace of septic replacements accelerated. In 2009, the CWC completed more than 360 septic repair and replacement projects, a 25 percent increase over 2008.[40] Its board of directors routinely devoted substantial portions of their monthly meetings to approving complex septic projects costing over $20,000.[41]

In addition to replacing individual septic systems, the CWC oversaw the construction of community septic systems in more densely populated villages. Community systems made economic and environmental sense in places with chronic septic problems. Rather than rebuild individual systems, many of which were located on unsuitable soils and could not accommodate the volume of waste generated by households, the CWC opted for community solutions in which waste from dozens of households was transported to an appropriately sized and located leaching field.

The installation of a community septic system in the Delaware County hamlet of Bovina Center delivered on the MOA's commitment to simultaneously protect water quality and promote economic development. The village's sensitive location—it sits astride the Little Delaware River, which empties into the Cannonsville Reservoir—prompted the CWC to undertake simultaneous septic and storm-water improvements. Settling tanks and a leach field purified wastewater from the village's seventy-five buildings; grassy swales, culverts, and catch basins reduced storm-water flows into the Little Delaware. CWC funds also paid for basic amenities, such as sidewalks, that the village had long coveted. Bovina residents appreciated the water quality improvements, but they were particularly delighted with the new roads and sidewalks. Village Councilman Chuck McIntosh enthused: "That's what's so great about this project. We have a sidewalk for those children to ride your bicycles, for their mothers to push the baby carriages. That has been a huge improvement in the village of Bovina. And I love to see young families come here, because that's what makes a community thrive—its young families."[42]

The all-encompassing nature of watershed protection made it difficult to attribute improvements in water quality to specific interventions such as septic replacements. Moreover, as the DEP observed of its septic, storm-water, and farm pollution efforts in 2006, "Because pathogen detection rates have always been low in the Catskill and Delaware watersheds, it is difficult to quantify the positive effects these programs have had on water quality[;] however based on how these programs perform, it is beyond dispute that any risk of potential pathogen contamination has been significantly reduced."[43]

Residents singled out septic repairs as an especially effective means of reducing microbial contamination in the city's water supply. Alan White emphasized the water quality benefits of connecting households with failing septic systems to community septic systems or wastewater treatment plants.[44] The CWC's

Rosa agreed. Septic and storm-water projects, he argued, had resulted in notice-able improvements to water quality. Although he acknowledged the difficulty of empirically demonstrating the effectiveness of these programs, Rosa cited his own experiences as a fly fisherman: "I have noticed these streams being clearer. I would have to say, they just seem to be fresher than they were twenty years ago. I can remember fishing up there just below Fleischmanns . . . and if you went down through there mid-July trying to fly-fish, several times you'd get a whiff of some-thing that you knew wasn't quite what it was supposed to be there. You don't get that now. . . . The water just seems a lot cleaner."[45]

From the perspective of Catskill residents, purer streams vindicated their insis-tence that environmental protection, community preservation, and economic de-velopment were compatible goals. New York's 1990 draft regulations had called for a five-hundred-foot buffer between septic systems and watercourses instead of the state standard of one hundred feet. Coalition attorney Jeff Baker observed that the five-hundred-foot rule "would have made enormous areas completely undevelop-able. It would have completely changed the pattern of how to do development and how to replace a system." Baker argued that the five-hundred-foot buffer was arbi-trary and reflected the ambitions of environmentalists, who viewed protecting the watershed as a useful vehicle for limiting development. He relished the memory of a 1992 presentation on septic systems and waste transport by engineers hired by the coalition. "DEP showed up, with like twenty-five or thirty people . . . and they were in rapt attention, learning from these guys basic hydrogeology. And I was sit-ting back there thinking—you know, I'd had my 'rocks for jocks' class in college—I know this stuff, they don't know this stuff? And they came up afterwards, they were asking lots of questions."[46]

The MOA balanced the city's valid concerns about the threat of septic contami-nation with the need for economic development. The city agreed to maintain the more flexible state standards and curb sewage contamination by funding a septic repair-and-replacement initiative. Just as in the farm program, judicious invest-ments in new infrastructure and solutions customized to community needs and variable soil conditions proved a more effective long-term solution than uniform regulations. By forcing the city to adopt a fiscal rather than a regulatory solution to the problem, watershed residents ensured that curbing septic contamination would not hinder economic growth.

Despite the popularity of the farm and wastewater programs, a new wave of discontent began to wash over the watershed in 2006. Alarmed at the prospect of continued large-scale land acquisition, Catskill residents reprised their strategy of the early 1990s and turned to the courts to limit city purchases. Once again, the future of New York's water system looked decidedly uncertain.

Acquiring Land and Unsettling the Neighbors

On the difficult issue of land acquisition, the MOA struck a delicate balance between the art of politics and the science of watershed management. Using the power of eminent domain, New York City had taken title to tens of thousands of acres of prime agricultural land to build its Catskill reservoir system. The resentment provoked by these land seizures lingered for generations. Pursuing a negotiated settlement to protect the watershed precluded the use of eminent domain. By making land purchases by the city contingent on the willing participation of Catskill property owners, the MOA charted a new course for watershed management in the United States. Although state-owned land in the Catskill Forest Preserve substantially enhanced water quality in the Ashokan and Pepacton watersheds, the state owned little land in some portions of the Catskills, and the city generally owned only a narrow band of land surrounding its reservoirs. As a result, public lands comprised less than a quarter of the west-of-Hudson watersheds in 1997.[47] Purchasing more of this "living landscape" raised a series of questions: Would watershed residents sell enough land to significantly affect water quality? Would the DEP be able to successfully manage these lands? How would city land acquisition affect the local economy and land values?

From the DEP's perspective, the biggest unknown was the quality of land it could potentially acquire. The MOA required the agency to devise a prioritization system for land acquisition based upon estimated travel time of water flows from a parcel to a reservoir. Properties whose flows emptied into city reservoirs most quickly and which drained near water-intake locations received top priority. The DEP also targeted land near reservoirs connected directly to the distribution system because water from these reservoirs would be the first to flow from kitchen and bathroom taps in the city. The DEP packaged these various considerations into a five-tier prioritization system that guided its solicitation decisions. MOA rules were designed to ensure that land acquisition reflected risks to water quality. For example, while the EPA required the city to solicit offers on virtually all land within the top two priority levels, it was mandated to make offers on only 50 percent of the lowest-priority lands.[48]

The performance of the land acquisition program (LAP) must be viewed in the larger context of the decision to implement an ambitious watershed management strategy instead of constructing a vastly more expensive filtration plant. This decision committed the city to a democratic process embodied by the willing seller / willing buyer provision; efficiency played little role in the design of the land acquisition program. Because efficiency and equity generally worked at cross-purposes, the holy grail of watershed protection became policy measures that would both enhance water quality and earn the support of watershed residents.[49]

In analyzing the LAP, then, it makes sense to begin by evaluating its effect on water quality. From 1997 to June 2009, the DEP acquired almost ninety-eight thousand acres in the Catskill system, roughly two-thirds through direct purchases and one-third through conservation easements that limited development. (The vast majority of these easements were obtained through the Watershed Agricultural Council's agricultural easement program.) These acquisitions increased the percentage of publicly owned land in the watershed from 24 percent to 34 percent. The DEP offered a surprisingly frank assessment of the effectiveness of these purchases: "Land acquisition is an anti-degradation tool that does not have any immediate impact on water quality. Further, it is impossible to predict with certainty whether or how a property protected by LAP might have been developed, and how such development would have impacted water quality." The MOA generally prohibited the DEP from purchasing property that contained homes and businesses; by definition, then, it acquired undeveloped land that posed little immediate threat to water quality.[50]

In some instances, land purchases actually degraded water quality. DEP employees were often dismayed to find that a landowner had thoroughly logged his or her property just before the city was scheduled to take possession. Because city land assessments did not include the value of a property's standing timber, they significantly understated the value of heavily forested properties. To realize the full value of their holdings, landowners logged the timber and then sold their ravaged properties to New York. As Alan White observed, "what we're getting is a bunch of . . . liquidation timber sales, prior to the liquidation of the property. And that's *bad* watershed management."[51] These sales of denuded properties clearly did not strike an acceptable compromise between the desire of residents to profit from their holdings and the need to protect water quality.

The DEP did achieve some notable successes in protecting particularly sensitive portions of the watershed. All water from the Delaware system flows through the Rondout Reservoir, which is connected by tunnels to the Cannonsville and Pepacton Reservoirs. The Delaware Aqueduct then conveys Rondout's waters to the West Branch Reservoir in Putnam County for delivery to New York City. Prior to the MOA, the city owned little buffer land around the Rondout, and relied primarily on state holdings to ensure water quality. Aggressive solicitation in sub-basins with streams flowing into the reservoir increased the percentage of protected land in these areas from 19 percent to more than 45 percent. Because land values in the Rondout area were among the lowest in the watershed, New York was able to purchase this protection at a relatively affordable price.[52]

The situation at the other end of the Delaware Aqueduct was quite different. Unlike the Rondout area, whose remote location largely insulated it from development pressure, in the 1990s the area around the West Branch Reservoir experienced a surge of new residential and commercial construction. The environmental organization Riverkeeper painted an alarming picture: "Like mushrooms after a rainstorm, shopping malls, office complexes, movie theaters, self-storage facilities, religious compounds, apartment buildings, manufacturing centers and sprawling housing developments are sprouting up in virtually every unoccupied corner of the watershed."[53] Purchasing extensive tracts of land in mountain watersheds was a necessary but not sufficient step; New York would also need to acquire sensitive lands in the Croton region to ensure that Catskill waters did not degrade before delivery to the city. The development pressures that made ensuring the purity of the Delaware waters in the West Branch Reservoir in Putnam County so difficult were even more intense near the Kensico Reservoir, the terminus of the Catskill Aqueduct.

The bloom of homes, parking lots, and corporate headquarters in the Croton watershed imposed a financial double whammy on the city. The primary solution to the decline in water quality—the Croton Filtration Plant—cost more than $4 billion. The secondary fix, purchasing land around the Croton reservoirs, also proved expensive. The willingness of several large landowners in the West Branch to sell to the city significantly boosted the percentage of protected land around the reservoir, but it came at a cost of almost $80 million, approximately one-third of total land acquisition spending in the first ten years of the program. Acquiring only two hundred acres of land near the heavily developed Kensico Reservoir, which had some of the highest suburban land values in the country at an estimated $250,000 an acre, consumed tens of millions of MOA dollars.[54]

Many residents of Putnam and Westchester Counties supported DEP land acquisition because they had grown weary of the onslaught of building in their communities. Their counterparts west of the Hudson were decidedly less enthusiastic. Resistance centered largely on the city's reluctance to open newly purchased lands for recreation. The dramatic easing of access restrictions initiated by Commissioner Lloyd in 2006 greatly reduced the friction over recreation. When the DEP announced in 2008 that it planned to make city-owned land adjacent to state property open for hunting, fishing, and hiking without a permit, Coalition of Watershed Towns chairman Dennis Lucas called the decision "further evidence of the strong partnership between the City and the people of the Catskills."[55] But only a year before Lucas made this statement, the coalition had sued the EPA to prevent the continuation of the land acquisition program. Conflict over land purchases threatened to unravel a decade of progress.

Ironically, many of the frustrations expressed by watershed residents could be traced to the willing seller / willing buyer provision intended to limit purchases. The city's inability to purchase land from unwilling owners ensured that its newly acquired properties were widely scattered across the west-of-Hudson watershed. Although the priority system was designed to preserve lands that provided high levels of water quality protection, over 65 percent of the acreage the city protected was located in the two lowest-priority areas. These parcels were generally located in the more remote and agricultural portions of the watershed where land was relatively inexpensive. Though this outcome was somewhat predictable—the two lowest-priority areas contain the most watershed acreage and thus could be expected to account for the majority of sales—it rankled Catskill residents.[56]

In the eyes of many residents, purchases in marginal portions of the watershed amounted to a land grab and recast the city in the familiar role of imperial intruder. After the DEP acquired two large tracts in the town of Harpersfield, almost sixty miles from the Cannonsville Reservoir, Delaware County supervisor Jim Eisel fumed, "there is no science here."[57] Acquisitions with little direct impact on water quality frustrated officials who viewed the MOA as a partnership in which residents agreed to curb pollution in exchange for infrastructure and economic development benefits. These purchases failed on both counts. They did little to enhance stream quality, and hampered economic growth by reducing the supply of potentially developable land. Moreover, because many of those who sold to the DEP were second-home owners from New York City, these transactions undermined the spirit of the willing seller / willing buyer provision. In these cases, watershed officials argued, the city was essentially buying land from itself.[58]

The combination of city land acquisitions and challenging natural conditions unnerved industry and government leaders eager to expand Delaware County's economy. Abundant streams and steep slopes rendered much of the land base unsuitable for development. DEP land purchases and easements near town centers in villages such as Hobart and Stamford further constrained the potential for new homes and businesses.[59] In this context, agricultural easements, once viewed as a classic win-win solution that would protect water quality and provide farmers much-needed cash infusions to continue operating, no longer seemed so benign. Nearly a quarter of the acres the DEP acquired in low-priority areas were in the form of agricultural easements.[60] Delaware County watershed coordinator Dean Frazier recounted a case in which an easement worked against another environmental goal in Stamford: "The farmer sold an easement to WAC. Well, that took some of the land that was identified in the community waste system area that they now could not use for their community waste. A few things like that happened. There was nothing nefarious about it; it was just the way it happened. But it created conflict."[61] It is worth underscoring Frazier's observation that city easements were not "nefarious." Despite a framework of complex rules and regulations and improved relations between the city and watershed interests, the possibility that New York's actions would stifle watershed economic development threatened to undermine the MOA. Maintaining an acceptable balance between source protection and economic vitality required frequent readjustments.

By early 2007, county officials realized that the fight to modify the land acquisition program was not a simple reprise of the classic city-vs.-country duel. The DEP had little appetite for continued large-scale land acquisition in the watershed. At a New York City Council meeting in January 2007 it proposed spending only $50 million over the next five years. Frazier attended the meeting and recalled the reaction the proposal generated: "EPA gets up and 'That's not enough; you've got to do more than that.' . . . DEC said that's not enough. And then there was a number of environmental groups there, six, eight, nine, ten of them, they all got to testify, and they're all saying this is not enough."[62] The filtration requirement and the MOA gave the state and federal governments substantial regulatory authority over New York City's water system. Watershed residents came to understand that the state could be a fickle ally. The same state agency that strongly supported increasing recreational access on city watershed lands now joined the chorus calling for an ambitious new round of land acquisition by New York. With the future of filtration avoidance in the balance, the DEP had no choice but to yield to state and federal pressure.[63]

The EPA's announcement in April 2007 that it intended to grant New York City a ten-year extension of its filtration waiver offered something for all stakeholders—except watershed residents. The waiver required the city to devote an additional $300 million to land acquisition over the next decade, which included $59 million remaining from previously allocated funds. It directed the DEP to take more concrete action to reduce turbidity in some supplies, and to construct an ultraviolet light disinfection plant by the summer of 2012 to reduce the risk of microbial contamination. The waiver also envisioned the continuation of the partnership programs through the Catskill Watershed Corporation and the Watershed Agricultural Council. Extending the waiver period to ten years satisfied New York City's demand for long-term certainty and reduced its annual acquisition expenditures while still guaranteeing the continuation of a vigorous land protection program.[64] Environmental lawyer Eric Goldstein articulated the conventional wisdom: "It represents a commitment among all of the parties—the city, state and federal government—to focus on the challenges of protecting the source water supply rather than pursue a costly and gargantuan construction project." Or, in tabloid parlance, "NYC Ducks $8B Soaking."[65]

Watershed residents were flabbergasted by the news. A decade of working closely with city officials to find common ground appeared to be for naught. During the MOA talks, state and federal officials played supervisory roles; municipal officials and watershed residents did most of the active negotiating. A decade later, watershed residents discovered they no longer had a place at the table. Under Governor

Pataki, the state recognized that a sustainable watershed program must have the support of Catskill residents. Under Governor Spitzer, satisfying the demands of environmental advocates appeared to have top priority. Dean Frazier recalled the FAD-renewal process: "There was no negotiation with anybody in this watershed, signatories to the MOA. We were essentially told that we don't count anymore. All communication with us stopped. When I say "us," the entire west of Hudson. The administration, environmental groups went behind closed doors. We've said this publicly, and nobody's denied it. They cut a deal."[66]

Watershed residents chose the only available option—a return to the familiar cycle of lawsuits and negotiations that had dominated the 1990s. The coalition sued the EPA to block implementation of the waiver. The Second Circuit Court of Appeals ruled that the coalition lacked standing, and dismissed the suit in late 2008.[67] The coalition then filed a second lawsuit alleging that the DEP's plan for land acquisition did not comply with the environmental impact requirements of the State Environmental Quality Review Act. This time around, the coalition was on firmer legal ground. Rather than take a losing case to court, city attorneys elected to negotiate. The coalition's main goal was to modify the land acquisition program to mitigate its negative economic impact. As the two sides made progress on the most contentious aspects of the LAP, city attorneys began to eye the calendar anxiously. The DEP's state water supply permit was due to expire in 2011. The permit represented an opportunity for Catskill residents to challenge the legitimacy of the entire watershed program. Rather than launch a second round of negotiations directly on the heels of the LAP discussions, the city decided to pursue a comprehensive agreement that addressed virtually all the issues dividing the two sides. In exchange, the coalition agreed to suspend its lawsuit and to refrain from challenging the water supply permit contingent on the success of the negotiations.[68]

Once again, watershed activists successfully deployed environmental laws to bring New York to the negotiating table. Their reliance on the state Environmental Quality Review Act to limit land acquisition was an ironic ploy; the law's intent was to require the disclosure of potential environmental consequences of government projects, not protect the development rights of small communities. New York's overriding concern was renewing its state water supply permit. Without it, the city could not continue to acquire land, and would lose its filtration waiver. The coalition's exclusion from the filtration renewal negotiations harked back to the pre-MOA era, when outside interests reshaped the Catskills with little regard for the welfare of local residents. But state and federal environmental laws gave watershed residents the clout they had once lacked. Legal challenges could undermine the watershed program, potentially forcing the city to construct a filtration plant. Significant cost overruns on the Croton filtration plant reinforced New York's desire to compromise with the coalition. In the ecological era, negotiating was simply much cheaper than attempting to maintain imperial rule.[69]

The negotiations that began in 2008 had much in common with the talks that culminated in the MOA. They dragged on for more than two years, covered a variety of issues, and involved dozens of people representing watershed residents, the city, the state and federal governments, and environmental organizations. Watershed residents negotiated with the benefit of a decade's hard-earned perspective. As coalition attorney Jeff Baker observed, the negotiations represented a "Halley's comet moment of influence" to make meaningful modifications to the most objectionable city practices.[70] The coalition could not substantially reduce the scale of land acquisition because the filtration waiver established spending targets and solicitation requirements. But it did succeed in modifying other aspects of the acquisition program. New York agreed to permit villages and towns to increase the size of areas designated as off-limits to city land acquisition. Although this concession removed only a few thousand acres from potential acquisition, it provided assurance to Catskill communities that city easements and purchases would not hinder growth in already developed areas. To address complaints that much of the land it acquired had minimal impact on water quality, New York agreed to provisions requiring newly purchased lands to have natural features, such as proximity to a watercourse and minimum steepness of slopes, important for water quality. These changes removed an additional twelve to fourteen thousand acres from potential acquisition.[71]

Negotiators also wrangled over the tax status of property New York had owned for decades. Agreeing to engage in comprehensive negotiations meant finally resolving the long-standing dispute over how to appraise the city's waterworks facilities. Under the terms of the MOA, the city was prohibited from challenging assessments on newly acquired parcels for twenty years from the date of purchase. Watershed residents insisted on this provision because New York routinely challenged its tax assessments. However, the city continued to dispute the assessed value of its dams, reservoirs, and sewage treatment plants. Small reductions in tax assessments on these high-value holdings could leave the budget of a small Catskill town in tatters.

Nearly a century after building the Ashokan Reservoir, the city continued to dispute the value of the property submerged by the reservoir. Jeff Baker recalled the dueling assessment strategies adopted by the towns and New York City:

> We'd have our appraiser and assessor go out and we'd find like farmlands or things that are in valleys, because that's where you're building the reservoir, and we came up with values of $5,000 an acre. . . . The city was coming up with values of a few hundred dollars an acre because their comparables they

were taking were on the top of mountains. And you know, it's like, yes, but you don't build the reservoirs on the tops of mountains.[72]

On the surface, the dispute was all about money. At a deeper level, the tax conflict rekindled long-simmering resentment over the city's appropriation of the most valuable farmlands in the Catskills. Reservoir construction dealt a heavy blow to Catskill agriculture because, as Baker noted, both activities required valley lands. When New York appraised reservoirs and dams as if they were located on remote mountaintops, it not only lowered their potential tax valuation but also distorted the historical record by implying that the soil they were built on was marginal, rather than central, to the Catskill economy.

Watershed residents and their attorneys insisted that any deal must resolve the tax status of high-value city watershed holdings. After decades of depleting their treasuries to fight New York's protests (most of which failed), municipal officials in the Catskills wanted some assurance that the future would be different. The city's penchant for challenging tax assessments had drained the $3 million in MOA funds set aside to cover the legal costs of watershed communities. After months of fruitless negotiations with attorneys from the Division of Budget, Baker appealed to the city's environmental lawyers to intercede. Pressure from these attorneys and the desire to reach an agreement before its existing water supply permit expired prompted New York to yield substantial ground on the tax issue. While the agreement did not set specific values for city-owned infrastructure, it did establish a template for determining property values. If a town chose to adopt this collaborative approach, New York was obligated to participate. The terms offered the prospect of financial stability for Catskill towns with significant reservoir acreage. The resolution of the tax issue paved the way for a comprehensive watershed pact, a sequel of sorts to the original MOA. In addition to modifying its land acquisition program and abiding by the revised taxation procedures, New York agreed to continue to fund the partnership programs, an important source of economic activity for the watershed region.[73]

In some respects, the agreement and the continued success of partnership programs marked a new phase in relations between the city and watershed interests. The instinctive mistrust that had long prevailed began to dissipate. New York continued its practice of increasing recreational opportunities on watershed holdings; from 2003 to 2011, the DEP more than doubled the acreage available for fishing, hiking, hunting, and other activities.[74] The strong performance of the CWC and WAC reassured municipal officials that these organizations acted in New York's best interests. City officials recognized that the local character of these organizations encouraged watershed residents to install new septic systems, sell agricultural easements, and take other steps to improve water quality. A 2006

DEP report concluded that "no protection program for the City's water supply, no matter how carefully crafted, can succeed without support and involvement of the City's partners and watershed stakeholders. Perhaps the greatest achievement of the past 15 years of has been the development of vital, locally-based organizations working with the DEP on the common goal of watershed protection."[75]

For their part, Catskill residents and officials began to accept that the city, its reservoirs, and the watershed program had become a permanent fixture of the political and physical landscape. As the CWC's Alan Rosa observed, "more and more people are realizing that it's a drinking water supply not just for the city of New York but half the population of the state of New York, and half the population of the state of New York is not going anywhere. But sixty-six thousand people in the watershed could go somewhere; that's the reality of it."[76]

With a fifteen-year state water permit finally in hand and a filtration waiver that lasted until 2017, DEP officials shifted their attention to upgrading the water system's sprawling network of aqueducts, tunnels, and dams. The most pressing challenge was repairing large cracks in the Delaware Aqueduct that leaked approximately thirty-five million gallons of water a day. The cracks developed in areas where the aqueduct passes through limestone, which erodes more quickly than sandstone, shale, and granite.[77] Plugging the leaks in the world's longest continuous tunnel would bring much-needed relief to the upstate towns of Wawarsing and Roseton, whose residents had become accustomed to habitual flooding of their yards and basements. It would also increase the flow of water to New York's hydrants and faucets by almost 5 percent, providing an extra cushion during droughts. The DEP first learned of the leaks in 1988, but took no action even as the estimates of water leakage increased from 15–20 MGD to well over 30 MGD. The state and environmental organizations sharply criticized the city for failing to adequately monitor and repair the leaks. The staid language of a 2007 state report could not obscure its alarming message: "If conditions . . . in the tunnel deteriorate further, the water supply for millions could be disrupted. However, we found that DEP does not have an adequate emergency response plan in place to address the sudden failure of the tunnel."[78] Flooding of residents' homes intensified beginning in 2008. One Wawarsing couple "reported that their concrete floor buckled from the pressure beneath it, and that even with a sump pump operating 24 hours a day the basement fills with 8 to 12 inches of water."[79] In 2011, the city and state ended the nightmare for many Wawarsing residents by agreeing to purchase their damaged homes.[80]

The DEP's failure to aggressively remediate the leaks stemmed largely from its insistence on viewing the problem as part of a comprehensive plan to improve the dependability of city supplies. During the years the DEP spent developing the plan, the magnitude of the leaks increased.[81] The greatest challenge associated with repairing the leaks was locating an alternative water source to replace the aqueduct's normal flows. Because the aqueduct carries approximately 50 percent of daily flows, the DEP cannot shut it off without providing for alternative supplies. Plans call for constructing a three-mile bypass tunnel around the most affected section near the Hudson River and making additional repairs to other failing segments of the aqueduct, most notably in the Wawarsing area. The bypass project, forecast for completion in 2020, will require the complete shutoff of the Delaware Aqueduct for up to a year. To compensate for the drastic reduction in supplies, the DEP will increase deliveries from Croton reservoirs and connect the Catskill and Delaware aqueducts to enable delivery of water from the Delaware reservoirs via the Catskill Aqueduct.[82]

Almost one hundred miles south of Wawarsing, crews labored hundreds of feet beneath Manhattan to complete City Water Tunnel No. 3, the longest-running and most expensive construction project in the city's history. Many of these men—known as sandhogs—have devoted their entire working lives to building the tunnel. When completed in 2020, the sixty-mile tunnel will deliver water to all five boroughs, allowing the DEP to close and inspect the other two delivery tunnels for the first time. In perhaps the riskiest wager in the city's history, DEP officials are betting that the sandhogs will finish City Water Tunnel No. 3 before the other tunnels fail.[83]

Yet another construction project, this one in Westchester, underscored the city's multifaceted approach to safeguarding water quality. As a condition of its ten-year filtration waiver, the DEP was required to construct an ultraviolet light facility to treat mountain water. Beaming mountain water with ultraviolet rays will reduce the risk of microbial contamination. At a cost of $1.6 billion, the price is steep, but still much less than constructing a plant to filter Catskill water. On the one hand, New York relied on trees, rocks, and land purchases to provide it with high-quality water; on the other hand, it deployed the latest technology to buffer its reliance on nature. This point bears repeating: watershed protection was not an anti-technological strategy. Rather, it dictated the use of particular technologies to complement its emphasis on pollution prevention.

It was a new technology—the use of hydraulic fracturing to extract natural gas—that most unnerved DEP officials. The possibility of natural gas drilling in the Catskill watershed represented the greatest external threat to the city's water supply. New technologies allowed energy companies to profitably extract the gas embedded in shale beds by injecting a chemical cocktail to fracture the shale (ergo "fracking"). The combination of horizontal drilling, whereby a single well can be turned horizontally to tap large veins of gas-bearing rock, and fracking made natural gas deposits previously off-limits economically viable. Energy companies had long been aware of substantial gas reservoirs embedded in the Marcellus Shale formation, which stretches across Pennsylvania and New York southwest into West

Virginia and west to Ohio. But it was only after gas companies demonstrated the feasibility of fracking in Texas, Wyoming, and other western states that they began to eye Marcellus Shale deposits.[84]

Few Northeasterners had ever heard of fracking in 2005. Yet by 2010, energy companies had drilled hundreds of wells in Pennsylvania and several dozen in New York State, and fracking had become the region's most contentious environmental issue. Fracking advocates hailed the Northeast's vast deposits as an economic and environmental boon. Tapping into this domestic source of energy, they argued, would enhance national security and stimulate rural economies. In addition, they claimed, natural gas was an environmentally responsible fuel and "burns cleaner than other fuel sources, with less pollutants and no mercury."[85] Opponents of drilling vigorously challenged these claims, arguing that the environmental consequences of drilling far outweighed the potential economic gains. Despite burning cleaner than coal, natural gas from fracking, they insisted, did not reduce overall greenhouse gas releases. Fracking's heavy transport footprint—fleets of heavy trucks were needed to deliver the water, chemicals, and other items required to sustain drilling operations—and inadvertent releases of methane in the extraction process generated large volumes of emissions, undoing its supposed environmental superiority.[86] Environmentalists also contended that fracking posed a grave threat to water supplies due to the potential migration of gas into aquifers, chemical spills, and the failure to adequately treat contaminated wastewater laden with radioactive materials.[87]

Opinions in the Catskills were also split. Some residents worried that the environmental repercussions would mar the landscape and damage their health. A handmade lawn sign on the outskirts of Delhi in May 2011 clearly conveyed these fears: "Frack = Cancer." The DEP strongly opposed drilling in the watershed, arguing that the injected fracking chemicals could penetrate underground streams and eventually find their way into the city's reservoirs. It also claimed that gas drilling would result in the deforestation of several thousand acres and generate enormous volumes of wastewater. In addition, massive subsurface disturbance could damage aqueducts already in need of major repairs.[88] In 2008, New York State declared an effective moratorium on new hydraulic fracturing permits to allow time for a detailed environmental impact assessment. The DEP hired engineers to study the likely impact of gas drilling and vigorously lobbied for a permanent de facto extension of the moratorium in the watershed. In a clear sign that alliances shifted depending on the issue, environmentalists and the EPA joined the city in opposing drilling. For their part, Catskill officials who supported gas drilling accused New York of exaggerating the potential negative impact of gas drilling in the watershed. Delaware County watershed coordinator Dean Frazier accused the DEP of pretending "Delaware County was flat," which enabled it to greatly inflate the number of potential wells and the corresponding volume of water and fracking fluid involved.[89]

The state's April 2010 decision neatly balanced upstate and downstate interests. It permitted horizontal drilling and fracturing under new regulations in most areas of the state, but established exceptionally stringent and cumbersome drilling requirements in the New York City and Syracuse watersheds. (Syracuse also does not filter its water supply.) Companies would be required to conduct separate environmental impact statements for each individual well in the watersheds, making hydraulic fracturing in these areas cost prohibitive. Although environmentalists continued to advocate for an outright ban and worried about the effect of drilling on water system infrastructure located outside the watershed, the state's decision effectively ended the outcry over possible contamination of New York City's water supply from gas drilling.[90]

Making Sense of Implementation

The effective ban on gas drilling confirmed what Catskill residents increasingly recognized: they and their neighbors were part of a unique political and environmental experiment that distinguished their region from the rest of the state and, in many ways, the rest of the nation. Only those residing in Boston's watersheds, a few hours northeast of the Catskills, could directly relate to their experiences. The trajectories of the water systems in New York and Boston are remarkably similar. Both cities established municipal supplies in the mid-nineteenth century, greatly expanded these systems in the twentieth century, and struggled to meet the legal and environmental challenges of maintaining an unfiltered water system at the turn of the millennium. Like New York, Boston launched a watershed management program to obtain federal waivers from filtration. Instead of negotiating terms of the watershed program with local residents, Boston relied primarily on a regulatory strategy. The Massachusetts Water Resources Authority, which operates the water system that serves Boston and dozens of surrounding communities, pushed through new laws regulating wetlands and stream setbacks. These laws, in combination with significantly higher percentages of protected land, allowed Boston to obtain filtration waivers without engaging in the intensive negotiations between municipal officials and watershed residents that reshaped the ecology and economy of the Catskills.[91] Although the cities took somewhat different approaches, they faced a similar set of incentives.

Both cities concluded that there were clear financial benefits to opting for watershed management over filtration. By 2017, New York will have spent nearly $2 billion protecting and managing upstate watersheds. Constructing a filtration plant for its Catskill supply would likely have cost at least four times that amount. Money is only part of the explanation, however.

In New York City's case, it is important to underscore the learning that took place through the implementation process. City officials began to search for areas of common ground, rather than seek to impose their own ideas. In some cases, this flexibility reflected the experiences of particular employees. For example, Paul Rush, an engineer who became a deputy commissioner under Emily Lloyd, grew up in the watershed and recognized the benefits of opening up city holdings for recreation. Instead of erecting signs urging residents to report polluters, the DEP began to invite residents to hunt and hike on its watershed holdings. If New York continued to irritate local officials by contesting tax valuations and acquiring land distant from reservoirs, at least residents now had forums in which to interact with city officials and, depending on the issue, might find support from the state government.[92]

The perspective of many Catskill residents had changed as well. Almost a century had passed since construction of the Ashokan Reservoir. In the Ashokan region and other parts of the watershed where memories of displacement had faded, people tended to harbor less animosity toward the city. In Delaware County, where older residents retained memories of reservoir construction and recalled stories of relatives forced to move, resentment lurked closer to the surface. But even in the villages and farms that surrounded the Pepacton and Cannonsville reservoirs, people recognized that the law provided them leverage in their dealings with the city that their grandparents did not possess. Officials saw no contradiction between condemning New York's actions in the local newspapers and sitting down with DEP officials to negotiate modifications to the MOA. The implementation process helped both sides hone their political skills.[93]

Finally, source protection did not entail radical changes to the landscape or in the division of natural resources. Watershed management meant preserving the landscape that had provided high-quality supplies for decades. Where farmers tilled the land and grazed their cattle, New York sought to maintain farms; where forests grew, New York sought to maintain forests through land acquisition. At its core, watershed management aimed to reduce water contamination by modifying the ways in which everyday activities were conducted. In some cases, such as farm operations, such changes were quite significant. In others, such as septic replacement, wastewater plant upgrades, and storm-water controls, watershed management meant improving on what residents were already doing to protect water quality.[94]

The results were politically innovative but ecologically conservative. The anti-environmental backlash of the 1980s encouraged public officials to devise more cooperative approaches to protecting natural resources. Policymakers and

analysts argued that collaboration and negotiation offered the best hope for preserving species and ecosystems. Within a decade, most of these well-intentioned
efforts had foundered, unable to find common ground acceptable to a wide range
of competing stakeholders. However, in the Catskills, where the traditional
upstate-downstate animosity was sharply magnified by successive rounds of waterworks construction and urban disregard for local prerogatives, the two sides
forged a landmark agreement and resolved the inevitable conflicts that arose during the implementation phase.[95] The MOA did not alter the basic ecological features of the region. The qualities that made the Catskills a highly desirable water
source to John Freeman and his fellow engineers at the turn of the twentieth
century—its thickening forests, dense network of creeks and streams, and low
year-round populations—remained largely intact in the early twenty-first century.
The dual thrusts of the MOA were minimizing the environmental impact of the
population and its infrastructure and acquiring land to protect future water quality, not remaking the landscape or reinventing the economic basis of everyday life.

The MOA's success was a testament to the thousands of people who devoted
their time to negotiations, changed their farm practices, attended meetings, and
installed new septic systems. It also confirmed the environmental and social logic
of both rural and urban life. The city needed partners to protect water quality, and
it found them in the Catskills. But too much rural growth would threaten this
supply. Maintaining a landscape of fields, forests, and streams dotted with small
villages and hamlets benefited both Catskill residents and those who relied on
their water.

Even as the DEP and Catskill residents struggled to implement the MOA,
urban living was increasingly seen as an environmentally responsible choice. In a
world where carbon emissions became the dominant ecological metric, the New
Yorker who lived in a small apartment and relied on public transportation suddenly became an environmental superstar. After several decades in which urban
areas were identified as the source of most of America's social ills, cities made
sense again. But urban life was only possible because the city had reached a productive détente with the Catskill residents who lived amid its reservoirs. New York
exchanged its money for some of the country's best water, and water bills increased
slightly as a result. It was the best deal New Yorkers had struck since they paid five
cents for a subway ride in the 1940s.[96]

Epilogue

Putting Politics in Its Place

For nearly two centuries, the New York City water system has reshaped natural and built environments. The reservoirs that dominate Catskill valleys are only the most prominent manifestation of this system. Standing on a street corner in Manhattan, a perceptive observer notices other components of the water network: wooden, spinning-top–shaped water tanks grace the roofs of thousands of medium-size buildings in the city, and hundreds of street-level water sampling stations allow DEP employees to test the purity of supplies throughout the five boroughs. Beneath the streets, unseen by pedestrians, a billion gallons of water a day courses through a complex network of tunnels, mains, and pipes.

In probing the political ecology of this complex system, I have emphasized the ways in which physical and the political are inextricably intertwined. Sociologists and political scientists have defined political ecology in various ways. From a historical perspective, the most useful definition is the relationship between political arrangements and environmental knowledge. At the most basic level, the evolution of New York City's water system reflects the shifting balance of political power in twentieth-century America. The city enjoyed substantial autonomy in constructing and operating its water network until the 1970s. With little resistance from state or federal authorities, it used its own funds to construct the nation's most extensive metropolitan water supply network. Once the state legislature approved the project in 1905, Catskill residents had no means to stop the destruction of their homes, farms, and communities.

The successful battle to modify releases from city reservoirs marked a decisive turning point in watershed relations. Since the 1970s, Catskill residents have shrewdly deployed environmental law to achieve their own goals. The increasing clout of watershed residents rested on a fundamental irony: they generally opposed environmental regulations, and insisted that they could care for streams, fields, and forests without interference from the government. Nevertheless, to portray those who live in the mountains as "antienvironmental" is to accept terms that make little sense in the world of practical politics. Watershed residents successfully resisted the stringent regulatory framework that New York attempted to impose in the early 1990s. Similarly, they do not share the vision of environmental organizations that wish to place the majority of land in the Catskills off-limits to development. But many of them have deep connections to their own fields and forests and to the public lands scattered throughout the region. By New York's own admission, their contributions have ensured the success of the MOA, enabling the city to avoid filtration. These results validate the socio-ecological premise of the agreement—that a living landscape where people farm, log, harvest bluestone, shop, hunt, hike, and fish can produce some of the country's highest-quality water.

One of the important lessons of the MOA is that public policy and, most significantly, public money, can greatly accelerate changes that were already taking place. Individuals and communities in the Catskills recognized the need to replace aging septic systems and update storm-water controls in the 1980s, before the city proposed its draft regulations. Without the hundreds of millions of dollars invested by New York, these changes would have occurred more slowly and on a much smaller scale.[1] This insight also applies to the massive toilet replacement program of the 1990s. In that case, the lure of public funds induced landlords to replace hundreds of thousands of toilets, the critical factor in slashing overall water use by one-quarter. Similarly, the adoption of new agricultural practices by hundreds of Catskill farmers substantially reduced the flow of pollutants into city reservoirs. These more environmentally sensitive and less capital-intensive techniques reflected nationwide agricultural trends. As WAC's Fred Huneke observed, the reversion to more traditional forms of agriculture was "not just occurring here, it's broader scale."[2]

But New York's pollution reduction initiative was notable in two main respects. The MOA was the most expensive and comprehensive watershed management program in the world. New York City invested more than $1.5 billion to curb pollution over more than two thousand square miles. This represented an enormous investment of time, money, and attention. The expanse of the watersheds and the imperative to strike a deal with local residents precluded an exclusively regulatory strategy. Simply revising outdated watershed regulations would not substantially enhance water quality. It was the establishment of a vigorous public-private partnership, its second distinguishing feature, that ultimately ensured its success.

Both of these developments occurred as a result of a traditional environmental regulation intended to ensure the safety of surface water supplies. Congress has frequently crafted environmental legislation that pushes private industry to develop the new technologies needed to comply with the standards established by the new law. Emissions standards that prompted automakers to develop the catalytic converter are a classic example of such technology-forcing legislation. In the case of the Surface Water Treatment Rule, legislators and the EPA were focused primarily on establishing minimum standards for filtered systems; the filtration avoidance provisions were included to satisfy the demands of the handful of cities that did not already filter their supplies. No one foresaw the vigorous private-public partnership that would ensue in New York's watersheds. Nevertheless, it is important to underscore the critical role of formal environmental laws and procedures. The Surface Water Treatment Rule, SEQRA, and the requirement that each water system in New York State obtain a water supply permit directly influenced the course of watershed management.

In evaluating the development of the city's water system, it is impossible to separate outcomes from political processes. Political processes—the mechanics of how decisions get made and resources are allocated—played an outsize role in determining winners and losers. From the commission hearings that resulted in paltry awards to some Catskill property owners to the Supreme Court cases that confirmed the city's right to tap Delaware waters, New York enjoyed distinct advantages over those who sought to restrict its ability to construct an expansive water network. As the environmental purview of the state and federal governments increased, the nature of political processes changed. In the 1950s, state authorities effectively rubber-stamped the Board of Water Supply's decision to bypass the Hudson in favor of tapping the Delaware. Four decades later, the city was compelled to participate in lengthy negotiations with watershed residents and satisfy the demands of state and federal regulators. The city's failure to adhere to state environmental impact statement guidelines provided the Coalition of Watershed Towns the leverage to extract significant concessions on land acquisition, taxes, and other disputes. More transparent political processes did, in fact, result in more democratic outcomes.[3]

Political processes had direct social and environmental consequences. The lack of adequate political checks in the 1950s resulted in the construction of the Cannonsville Reservoir, which submerged farms, communities, and forests. More rigorous evaluation of the project would have revealed the presence of substantial agricultural pollution, pollution that prompted New York to minimize its diversions from the Cannonsville in favor of purer supplies from other reservoirs. The city could likely have obtained as much water by reducing leakage in its delivery system and inside residences as it did by building the Cannonsville. The story of the dueling plans to build Cannonsville or pump water from the Hudson highlights the importance of political processes in shaping environmental outcomes. We cannot appreciate the radical shift in power relations that occurred in the waning decades of the twentieth century without understanding the narrow confines of the water supply debate in the 1950s.

This shift began in the 1970s, when resistance by Catskill anglers forced New York to alter its management practices to protect the health of aquatic ecosystems and the interests of anglers and canoeists. In the 1990s, rural residents used their power to negotiate the MOA. The MOA reduced pollution from roads, farms, and homes, concentrated development in existing population centers, and enhanced biodiversity by protecting both agricultural and forested areas. Curbing water pollution had a positive spillover effect on other aspects of the natural environment.

Despite the apparent success of the MOA and affiliated farm programs in striking a balance between economic development and environmental preservation, those most intimately involved with the programs recognized the limitations. The agreement could not stem the tide of disappearing dairy farms, a decades-long trend. Nor did it lead to the construction of significant tourism infrastructure needed to make the Catskills more attractive to visitors.[4] Nonetheless, in international environmental circles the MOA was soon hailed as one of the world's more important payment-for-eco-services success stories. The notion of paying local people to maintain ecological resources that benefit those from afar has become a staple of environmental policy. Although the 2010 international climate negotiations in Copenhagen failed to produce a major accord, negotiators agreed on the basic outlines of a worldwide Reducing Emissions from Deforestation and Forest Degradation Program (REDD) under which wealthier countries would fund forest preservation in the developing world.[5] As Fred Huneke recalled, those involved in watershed programs did not recognize that their work embodied the principles of this new environmental model: "And we didn't realize, I didn't realize it at the time, we've just got our nose to the grindstone and we're just kind of plugging away . . . and all of a sudden, you know, we realize, oh, that's what we were doing." This belated recognition reflected the organic nature of the watershed protection programs, which took shape and evolved in response to actual conditions. For example, when land acquisition appeared to become untethered from the goal of protecting water quality, Catskill residents intervened to limit the program's scope.[6]

By 2008, WAC and other organizations regularly hosted environmental officials from the developing world eager to learn more about this watershed success story. These visitors were impressed by the individualized approach to farm management at the heart of the Whole Farm Program and also by new infrastructure features such as massive manure storage sheds.[7] These officials were quick to realize, however, that they would not have access to New York City's seemingly bottomless coffers. They were most interested in low-cost interventions, such as streamside fencing to keep cattle out of water sources, which they could feasibly implement. In other cases, New York's source protection approach had little relevance to the much more challenging and widespread pollution problems of the developing world.

WAC communication director Tara Collins recalled the reaction of a group of Indian officials: "We took one group out and they were like, 'This is all great that you're starting at basically the headwaters and keeping it clean from point A. But we have so many polluters and such big issues all the way down.'"[8]

If New York's style of watershed management cannot be exported wholesale to the developing world, it nonetheless offers useful insights into the potential of adaptive governance and payment for ecosystem services. The success of the Whole Farm Program affirms the wisdom of one of the central tenets of adaptive governance: the importance of local knowledge. The reliance on locally based experts to provide customized advice greatly enhanced the credibility of the program in the eyes of Catskill farmers. New York City wrote the checks, but it wisely turned over program implementation to people who already enjoyed a good working relationship with local farmers. Most farmers had a strong environmental ethic and were receptive to altering practices to reduce water contamination from manure and farm chemicals, especially if the DEP paid for transition costs and if the changes promised to improve their bottom lines.

The willingness to spend billions of dollars to preserve ecosystem services distinguished the city's watershed management efforts. In New York's case, the enormous financial discrepancy between protecting its headwaters and building a filtration plant made the ecosystem services approach an obvious choice. However, the choices facing policymakers are rarely so stark. Investing in water quality, healthy forests, and productive soils makes environmental and social sense. Cleaning up pollution is almost always more expensive than preventing it. Pollution prevention can provide income for those who live in sensitive ecosystems, giving them an incentive to care for their natural environment. But local people need partners, and these partners often need an unambiguous incentive to preserve ecosystem services. In the absence of a legal framework that provides these incentives, such programs will remain the exception rather than the rule.

New York's success also underscores the importance of the larger socio-ecological context. The resentment provoked by decades of reservoir construction and continuing tensions over highway maintenance, taxes, and other issues made earning the trust of watershed residents the greatest obstacle to reaching an agreement. Although disputes arose concerning septic setbacks and access to city watershed lands, many of the most contentious issues were related to money and the effect of the agreement on the region's economic potential. Most notably, in this region of abundant precipitation, there was virtually no disagreement about the water itself—there was plenty to go around. The focus was on preventing pollution, and, except in occasional emergencies, not on allocating a scarce resource among a large group of stakeholders, a common scenario in California and other parts of the American West. Having a favorable natural environment narrowed the scope of conflict. In places where multiple players lay claim to a resource and have battled each other for decades to legitimize these claims, stakeholders must resolve both social and ecological conflicts, an extremely tall order. Formal laws and regulations may provide little guidance in forging creative solutions.[9]

But in the Catskills, dramatic legal shifts encouraged watershed residents to set aside their grievances and strike a deal with the city. Eastern water law and weak state and federal environmental controls gave New York access to the most productive streams in the Catskills. The Supreme Court required only modest concessions from the city in exchange for the right to construct reservoirs that reshaped the ecology and economy of the region. The sharp contrast between these early modest concessions and the significant financial and political compromises New York made to secure the MOA is striking. The establishment of modern environmental law dramatically altered the legal landscape. The determined resistance of Catskill residents and the threat of filtration knocked the city off its high perch and into a seat at the negotiating table.

The empire of water did not crumble. Catskill residents did not gain independence from New York; in fact, a source protection strategy only more firmly intertwined the fates of country and city. New York relied on Catskill residents to reduce pollution, and in exchange it agreed to maintain the region's economic vitality. This strategy worked in part because organizations like the WAC and the CWC mediated between urban and rural interests. City officials recognized the importance of putting a local face on efforts to clean up the watershed. The city did not surrender so much as stage a strategic retreat.[10]

As Christopher McGrory Klyza and David Sousa have observed, reconciling the welter of conflicting environmental laws and customs that accumulated over the past century and a quarter creates significant challenges for environmental policymakers. The Catskill water system is a product of the conservation era, when influential Americans believed in harnessing the fruits of nature for the benefit of society. In New York's case, the era lingered for several decades, with new reservoirs gradually coming online until the 1960s. The environmental revolution of the 1970s emphasized ecology, transparency, citizen participation, and government regulation of industrial pollution. Political scientist Shep Melnick captured the shift in environmental policy that occurred in this era: "It is hard to imagine, for example, that the Hetch Hetchy reservoir could have been built after 1970. John Muir would have quickly filed suit under the National Environmental Policy Act or Endangered Species Act. Pinchot would still be working on the environmental impact statement."[11] In the 1990s, in response to property rights advocates and charges that overly burdensome regulations stifled economic growth, the federal government sought to promote the need for dialogue and cooperation to solve environmental disputes. At the same time, it began to shift its focus to managing entire ecosystems rather than particular sources of pollution.[12]

The watershed agreement allowed New York to maintain its conservation empire, but under distinctly postcolonial terms. It was only when Congress amended the Safe Drinking Water Act in 1986 that New York was forced to get serious about protecting its watershed. The city's draft regulations reflected the standard, top-down governmental response to pollution. In response, watershed residents protested that the regulations violated their property rights. This sequence of events neatly embodied the nation's larger environmental trajectory. The MOA suggested that it was possible to strike a balance between property rights and environmental protection, and to harmonize the conflicting demands of several environmental eras. Instead of relying exclusively on a 1970s-style regulatory solution, the MOA combined regulations, land acquisition, technological upgrades, and changes in land management to protect water quality. A second round of lawsuits and negotiations from 2007 to 2010 underscored the fragility of the agreement. Of course, collaboration and dialogue are not enough to resolve many environmental disputes. But the willingness of Catskill residents and city officials to embark on the world's most expensive and ambitious watershed management program after almost a century of bitter conflict offered hope that environmental solutions are not as elusive as they may seem.

It is not surprising that an infrastructure system as extensive as the New York City water supply network had a profound effect on the geography and ecology of the watershed regions. The development of the water system ramified across southeastern New York State in unpredictable ways. Perhaps the most significant environmental legacy of the system was the recreational network it spawned. From Long Island to New York City to Westchester and the Catskills, the elaboration of the city's water network resulted in the development of a remarkable array of recreational assets. These parks, reservoirs, and forests have provided urban, suburban, and rural residents an invaluable resource in one of the most densely populated regions of the country.

The incidental quality of this recreational legacy does not diminish its significance. Rather, it highlights the diverse benefits that water networks often confer on their users. Historians have traditionally focused on the public health benefits that accompanied the establishment of municipal water supplies. The transformation of the simple act of consuming a glass of water from a risky proposition to the most mundane of daily activities is a miracle we should not take for granted. But, at least in the case of large systems such as the one constructed by New York City, water networks did more than extend life spans and improve health. They also shaped the built, political, and physical environment over a broad swath of territory.

Traces of this legacy were clearly visible in the early twenty-first century. A plaque commemorating the Croton Distributing Reservoir greets commuters in a subway passageway near the library. A couple of miles uptown a visitor will find the upper reservoir of Central Park, one of the park's most pleasant sections. In Westchester County, many walkers and joggers gravitate to the wide trail of the Old Croton Aqueduct, where "OCA" signs dot the landscape. In 2013, New York City is scheduled to complete the restoration of the High Bridge, enabling city residents to walk the Old Croton Aqueduct Trail from Manhattan to the Bronx and into Westchester. Despite its nominal completion in 1967, the water system's legacy is constantly evolving as its guardians strive to update it for the future.

Reflecting on twenty years' involvement with the city's water system, Fred Huneke dubbed the relationship between New York and watershed residents a "marriage of necessity." In this book I have explored the complex and often unanticipated social, ecological, and recreational offspring of this marriage. In the Catskills, waterworks construction and operation submerged thousands of acres of productive farmlands, displaced residents, ended the boardinghouse economy, and altered stream ecology. In time, watershed residents came to appreciate some of the benefits that flowed from hosting this network, such as fishing in the reservoirs, limits on strip mall development, and, in more recent years, extensive hunting and hiking opportunities on city lands.

The view from the city has changed as well. Nearly a century after staging a play that depicted an Indian chief graciously bestowing Catskill waters on New York, the naïve romanticism of waterworks expansion has vanished. Allegorical pageants have given way to a website trumpeting the construction of a new wastewater treatment plant in the Catskills and the acquisition of additional parcels by the DEP.[13] Clean water is no longer a free gift from nature, but rather a shared resource that can be preserved only through judicious investments and active engagement between the city and watershed residents. Pragmatism had become the watchword of both sides. As Assistant Commissioner of Water Supply David Warne observed in 2009, "We've got to figure out a way to make this work, we're all in this for the long term."[14]

Notes

Introduction

1. New York City's water network in the Catskill Mountains consists of two water systems, the Catskill and the Delaware. New York City completed the Catskill system in 1928, when it finished building the Schoharie Reservoir. Construction of the Delaware system began in the 1930s and continued until the late 1960s. Since the late 1960s, most work on the Catskill network has consisted of repairs or projects to create redundancy in the system, such as City Water Tunnel No. 3. The tunnel has been under construction, on and off, since 1970 and is forecast to be completed in 2020. Except where the context clearly indicates otherwise, "Catskill" refers to all New York City waterworks infrastructure and property in the Catskill Mountains, not the Catskill portion of the mountain water network.

2. For insight into the wide-ranging public health and environmental implication of waterworks construction, see Melosi, *Sanitary City,* and Blake, *Water for the Cities.* The timing of delivery of safe drinking water supplies varied both across cities and even within them. Filtration dramatically reduced the incidence of typhoid and other waterborne diseases. See Condran, Williams, and Cheney, "Decline in Mortality in Philadelphia from 1870 to 1930."

3. "Hydrological commons" comes from Fiege, "Private Property and the Ecological Commons."

4. This account is drawn largely from Koeppel, "Rise to Croton." For a detailed look at the creation of the Croton system, see Koeppel, *Water for Gotham.*

5. Nonetheless, it is important to note that because residents were required to pay for the cost of connecting to the system, many New Yorkers went without Croton water for years.

6. Philip Hone, a prominent businessman and former mayor of New York, penned the classic eyewitness account of the introduction of Croton water. See Nevins, *Diary of Philip Hone,* 624–26.

7. Morris, *Croton Ode.*

8. Galusha, *Liquid Assets,* 266.

9. These cities developed long-distance water sources during this same period. On Boston, see Elkind, *Bay Cities and Water Politics,* and Nesson, *Great Waters.* The literature

on water supply in Los Angeles and San Francisco is extensive. On Los Angeles, see Kahrl, *Water and Power*. On San Francisco, see Righter, *Battle over Hetch Hetchy*, and Brechin, *Imperial San Francisco*. For a work that examines both cities in the larger context of the state's water challenges, see Hundley, *Great Thirst*. See also Melosi, *Precious Commodity*.

10. On the Great Lawn, see Blackmar and Rosenzweig, *Park and the People*. On Long Island, see Caro, *Power Broker*. On the Old Croton Trail, see Cooper, *Walker's Guide to the Old Croton Aqueduct*.

11. Perry Shelton, interview by Nancy Burnett, February 20, 1996, 157 (part of *Behind the Scenes: The Inside Story of the Watershed Agreement*, a series of twelve oral interviews; hereafter cited as *BTS*).

12. For background on the MOA, see Platt, Hoffman, and Gregory, "Full Clean Glass?"; National Research Council, *Watershed Management for Potable Water Supply*.

13. Anthony Bucca, interview by Nancy Burnett and Virginia Scheer, February 15, 1996, *BTS*, 42.

14. National Research Council, Committee on Assessing and Valuing the Services of Aquatic and Related Terrestrial Ecosystems, *Valuing Ecosystem Services*; Pires, "Watershed Protection for a World City." On ecosystem services more generally, see Alcamo and Bennett, *Ecosystems and Human Well-Being*; and Bateman, Irwin, and Ranganathan, *Restoring Nature's Capital*.

15. National Research Council, Committee to Review the New York City Watershed Management Strategy, *Watershed Management for Potable Water Supply*; National Research Council, Committee on Watershed Management, *New Strategies for America's Watersheds*; United Nations, World Water Assessment Program, *Water for People*, 177–78.

16. See, for example, Andrews, *Managing the Environment*; Sussman, Daynes, and West, *American Politics and the Environment*; Kamieniecki and Kraft, *Business and Environmental Policy*.

17. The presence of many stakeholders and the recognition of the value of local knowledge are the defining features of this new collaborative approach. See Sabatier, Weible, and Ficker, "Eras of Water Management"; Brunner et al., *Adaptive Governance*, vi–xi; John, *Civic Environmentalism*; and Wondolleck and Yaffee, *Making Collaboration Work*. For a work that offers several concrete examples of collaboration between unlikely bedfellows, including a summary of the New York City watershed negotiations, see Daily and Ellison, *New Economy of Nature*.

18. Gordon, "City Tunnel, Catskill Water," 121.

19. Gleick, "Changing Water Paradigm"; Brooks and Brandes, "Why a Water Soft Path?"

1. From Croton to Catskill

1. "From Lake to Reservoir," *New York Tribune*, July 16, 1890; "The New Aqueduct Opened," *New York Times*, July 16, 1890.

2. Stern, Mellins, and Fishman, *New York 1880*, 536–38.

3. Teaford, *Unheralded Triumph*, 225; Levine, "In Gotham's Shadow," 19.

4. *Brooklyn Daily Eagle*, June 29, 1985, quoted in Levine, "In Gotham's Shadow," 66.

5. Proponents of consolidation argued that state law limiting municipal debt to 10 percent of a city's property valuation made it impossible for Brooklyn to raise the capital needed to modernize and expand its water system. Frank Bailey, a prominent figure in

Brooklyn real estate, argued that "the only relief seems to be in annexation to New York." See "Shall the Cities Wed?" *Brooklyn Daily Eagle*, October 4, 1894.

6. Editorial, *New York Times*, September 26, 1891.

7. Allen, Duane, and Fteley, *Report to the Aqueduct Commissioners*, 81. Daily per capita consumption in New York City almost doubled from 1890 to 1894.

8. Parrott, "Water Supply for New York City."

9. For historical consumption data, see Judd, *How Much Is Enough?* 27.

10. Weidner, *Water for a City*, 145.

11. Eugene McLean, *Report to Bird Coler* (no title indicated), MC 51, box 101, folder Correspondence August–September 1899, 16, John Ripley Freeman Papers (hereafter cited as JRF), Massachusetts Institute of Technology, Institute Archives and Special Collections (hereafter cited as MIT Archives).

12. These reports are contained in box 101, folder Correspondence August–September 1899, JRF, MIT Archives.

13. "Roosevelt May Act in Ramapo Deal," *New York World*, November 20, 1899, morning edition.

14. New York State Legislature, Assembly Cities Committee, *Brief on Behalf of the Ramapo Water Company*, 25.

15. "Ramapo Raid Would Rob New York City of $195,460,070," *New York Herald*, August 20, 1900.

16. Open letter from the Merchants' Association, "To the Taxpayers and Citizens of New York" (November 27, 1899), box 102, folder 6, Corr. 1899–1904, March 1900, folder 2 of 3, JRF, MIT Archives.

17. Skepticism flowered around the turn of the century. In his study of the press in this period, Richard Kaplan observes, "A permanent attitude of suspicion guided the modern reader." See Kaplan, *Politics and the American Press*, 166.

18. William King to Benjamin Odell, September 7, 1900, from Merchants' Association of New York, *Attitude of Candidates for the Legislature toward the Ramapo Water*, 7.

19. Ibid.

20. New York State Legislature, Assembly Cities Committee, Brief on Behalf of the Ramapo Water Company, 3.

21. Weidner, *Water for a City*, 148.

22. Eisenhut, *Brooklyn Water Supply*, 36.

23. E. J. Lederle to Commissioner of Water Supply John Oakley, October 30, 1905, Mayor George McClellan Papers, box 61, folder 626, Municipal Archives of the City of New York (hereafter cited as MACNY).

24. Freeman, *Report upon New York's Water Supply*.

25. Burr, Hering, and Freeman, *Report of the Commission on Additional Water Supply*, 6.

26. Ibid., 72.

27. Freeman, *Report upon New York's Water Supply*, 428. Historian David Stradling also noted the change in opinion regarding the suitability of Catskill water sources. See Stradling, *Making Mountains*, 149.

28. George Tauber to John Ripley Freeman, November 5 and 7, 1899, box 101, folder NYC Water Supply, 1899–1904 Correspondence, October 1899, JRF, MIT Archives.

29. Burr, Hering, and Freeman, *Report of the Commission on Additional Water Supply*, 233.

30. See Eastman, *Water Supply of New York*.

31. Burr, Hering, and Freeman, *Report of the Commission on Additional Water Supply*, 502.

32. "Examination of New York and Passaic Waters," Manuscript Collection 465, box 1, folder 1909–11, Desmond Fitzgerald Papers, MIT Archives.

33. Burr, Hering, and Freeman, *Report of the Commission on Additional Water Supply*, 558.

34. Ibid., 553.

35. Ibid., 54.

36. Ibid., 558.

37. The bill did not explicitly mention the Catskill sources, but it was widely understood that New York City intended to tap the Esopus Creek as its principal new water source.

38. Galusha, *Liquid Assets*, 93; "Dutchess County Water Rights," *New York Tribune*, April 28, 1904.

39. John Ripley Freeman to Henry Towne, October 31, 1904, box 103, folder Correspondence June–December 1904, JRF, MIT Archives. The *New York Tribune* came to a similar conclusion: "It has subjected the people up there to a vast amount of needless inconvenience in a stupid and arbitrary fashion." See "The City and Its Water," May 21, 1904.

40. Robert Grier Monroe to John Ripley Freeman, April 15, 1904, box 103, folder Correspondence January–May 1904, JRF, MIT Archives.

41. Galusha, *Liquid Assets*, 45–48; the most complete history of the region is Evers, *Catskills*.

42. Evers, *Catskills*, 334. Pure water ensured that hides did not rot; lime was used to remove hair from hides.

43. Ibid., 338.

44. Ibid.

45. Carpenter, "Report to the State Forest Commission," reprinted in Van Valkenburgh and Olney, *Catskill Park*, 177.

46. Carpenter, "Report to the State Forest Commission," 157.

47. Kudish, *Catskill Forest*, 64–71.

48. Parrott, "Water Supply for New York City."

49. Walter Sears to Commission on Additional Water Supply, August 5, 1903, box 102, folder Correspondence August 1903, JRF, MIT Archives.

50. For background on the establishment of the Catskill Forest Preserve and Park, see Van Valkenburgh and Olney, *Catskill Park*, 27–61.

51. Walter Sears to William Burr, April 8, 1903, box 102, folder Correspondence January–April 1903, JRF, MIT Archives.

52. Walter Sears to William Burr, April 15, 1903, box 102, folder Correspondence January–April 1903, JRF, MIT Archives.

53. Walter Sears to William Burr, July 2, 1903, box 102, folder Correspondence July 1903, JRF, MIT Archives.

54. "Citizens Oppose M'Clellan Bill," *Kingston Daily Freeman*, February 29, 1905.

55. "Board of Trade Tables Resolution," *Kingston Daily Freeman*, February 15, 1905.

56. Ibid.

57. Letter to the editor from "A Member of the Board of Trade," *Kingston Daily Freeman*, February 20, 1905.

58. "Board of Trade Tables Resolution," *Kingston Daily Freeman*, February 15, 1905.

59. "Citizens Oppose M'Clellan Bill," *Kingston Daily Freeman*, February 29, 1905.

60. "The Water Problem," *Kingston Daily Freeman*, February 21, 1905.

61. On reform in New York City, see Cerillo, "Impact of Reform Ideology."

62. Higgins, *Public Papers of Frank W. Higgins*, 38.

63. Undated and unaddressed memo by George Bliss Agnew, Astor, Lenox and Tilden Foundations, box 5, folder 2, George Bliss Agnew Papers, Manuscripts and Archives Division, New York Public Library.

64. Draft of Senate bill 1174, March 8, 1905, box 1, folder 8, George Bliss Agnew Papers, Manuscripts and Archives Division, New York Public Library.

65. For a conventional account, see Weidner, *Water for a City*, 176–77.

66. "No Sympathy for Ulster," *New York Tribune*, March 25, 1905.

67. "Mayor Pleads for Water," *New York Tribune*, February 22, 1905.

68. "The Development of New Rochelle," *New Rochelle Pioneer*, May 13, 1905.

69. Forstall, *New York, Population of Counties by Decennial Census*; "To Develop Westchester," *New York Times*, January 6, 1905; Works Progress Administration, *Historical Development of Westchester County: A Chronology*, vol. 2. See entry for June 1905, which discusses the economic benefits of rail construction.

70. "More Water," *Westchester News*, June 10, 1905.

71. Comment on "In the Croton Watershed," *Mount Kisco Recorder*, January 27, 1905.

72. "To Cut Off Water Supply," *New York Tribune*, January 12, 1905.

73. The corporation counsel manages New York City's legal affairs.

74. "Agree on Water System," *New York Tribune*, March 9, 1905; "White Plains Can Tap New York City Water Supply," *Westchester News*, March 1, 1905.

75. On the opposition of northern Westchester to the water pact with New York, see "Westchester May Compromise," *Mount Kisco Recorder*, February 10, 1905.

76. "Fight in Ulster County's Behalf," *Kingston Daily Freeman*, April 22, 1905.

77. Carrott, *Egyptian Revival*, 107.

78. *Bulletin of the New York Public Library, Astor, Lenox and Tilden Foundation*, June 1915, 487, contained in New York Public Library, Astor, Lenox and Tilden Foundations, box 46, folder 37, John Shaw Billings Records, Manuscripts and Archives Division, New York Public Library.

79. Doctorow, *Waterworks*, 57.

80. On Fifth Avenue, see Boyer, *Manhattan Manners*, 45–51, 184–91; Page, *Creative Destruction of Manhattan*, 21–68.

81. Butler, *In Favor of Retaining the Present Murray Hill Reservoir*, 3.

82. "New Public Library: Fine Point Must Be Decided," *Brooklyn Daily Eagle*, December 27, 1900.

83. Butler, *In Favor of Retaining the Present Murray Hill Reservoir*, 13.

84. Reed, *New York Public Library*, 5–6.

85. "Address to the Mayor, Aldermen and Commonalty of the City of New York, Presented by Board of Trustees to Honorable William Strong at City Hall," March 25, 1896, box 34, folder 3, John Shaw Billings Records, Manuscripts and Archives Division, New York Public Library.

86. "Old Columbia as a Site," *New York Times*, March 29, 1895.

87. "The Free Public Library," *The Independent . . . Devoted to the Consideration of Politics, Social and Economic Tendencies, History, Literature, and the Arts*, April 2, 1896, 13. On the desire to match the city-building achievements of European dictators and the embrace of refinement and display, see Scobey, *Empire City*, 187–92. Replacing the reservoir with a stately neoclassical library fit the goal of "converting ugliness to beauty." Wilson, *City Beautiful Movement*, 4.

88. Board of Aldermen and Common Council of the City of New York, Committee on County Affairs, *In the Matter of Proposed Removal of 42nd Street Reservoir*, 2.

89. Ibid., 19.

90. Ibid., 12.

91. "Reservoir Site Favored," *New York Tribune*, March 26, 1896.

92. Wallace and Burrows, *Gotham*, 1067–68.

93. Barnard, "New Croton Aqueduct."

94. Jackson and Dunbar, *Empire City*, 207–8.

95. Smyth, "New York's Great New Library," 518.

96. As American environmental history expanded its focus to include areas outside the West, conflicts over water use in the East have received more attention. See Stradling, *Making Mountains*, 140–76; Pisani, "Beyond the Hundredth Meridian," 466–82. The secondary literature on water in the American West is vast. Important works include Kahrl, *Water and Power*; Hundley, *Great Thirst*; Righter, *Battle over Hetch Hetchy*.

97. Nesson, *Great Waters*.

98. The classic study of suburban development in Boston is Warner, *Streetcar Suburbs*. On Westchester County, see Panetta, *Westchester*.

99. Designation of large swaths of the rural landscape as forest preserve and parkland generated intense friction with local residents who had become accustomed to drawing on the abundant resources of these areas. On the Adirondacks, see Jacoby, *Crimes against Nature*.

2. Up Country

1. Scott, *Seeing Like a State*, 184–90. Historians of urban water systems have frequently invoked the imperial metaphor. See Brechin, *Imperial San Francisco*. In his examination of the ties between New York City and the Catskills, David Stradling argues that the relationship was too complex to merit the "imperial" label. However, he notes that "if one wanted to describe an imperial relationship between the city and the country—between New York and the Catskills—the best story to tell would be that of the water supply." Stradling, *Making Mountains*, 14.

2. Paul Josephson highlights the ability of democratic governments to mitigate the negative consequences of massive infrastructure projects. See Josephson, *Resources under Regimes*. James Scott also recognized the self-correcting mechanisms inherent in democratic capitalism, namely the respect for institutions and the belief in private political economy. See Scott, *Seeing Like a State*, 101–2.

3. George McClellan, *A Tribute to the Engineers of the BWS of the City of New York*, box 104, folder NYC Board of Water Supply, 1921–22, JRF, MIT Archives.

4. Ibid. John Watson, a New York City printer who vacationed in the Catskills, mocked McClellan's pretensions in a handwritten caption of a 1910 photograph he took of the project's triangulation tower: "This is the only completed work on the Ashokan enterprise. George B McClellan took no chances of being forgotten in this connection; hence this huge pile was erected and an inscription bearing the name of the modest ex-mayor was engraved thereon even before the more practical work of building the dam was begun." See Town of Olive Archives Special Collections, box U, folder John Watson Photograph Collection 1910 (hereafter cited as TOA). Fittingly, the McClellan Tower was renamed the J. Waldo Smith Memorial in a 1936 ceremony. See box GVS, notebook no. 26, Vera Sickler Papers, TOA.

5. *New York World*, September 6, 1907; Galusha, *Liquid Assets*, 94–95.

6. *New York World*, September 6, 1907.

7. *Board of Water Supply Songs*, 1909. Found in box 104, folder Corr. Sept.–Dec. 1909, JRF, MIT Archives.

8. Ridgway, *Robert Ridgway*, 181.

9. Ibid., 196.

10. "Personals," *Catskill Water System News*, August 5, 1911, 16.

11. Ibid., November 5, 1912, 136.

12. "Reservoir Field Day," *Catskill Water System News*, August 5, 1913, 206.

13. *Board of Water Supply Songs*, 1909.

14. For background on welfare capitalism, see Dawley, *Struggles for Justice*, 158–59.

15. Ridgway, *Robert Ridgway*, 193.

16. Board of Water Supply of the City of New York (hereafter Board of Water Supply), *Catskill Water Supply*, 1915 ed.

17. All postcards are contained in the Chet Lyons Collection, Olive Free Library Archives (OFLA); photographs are from Barry Knight, "The Catskill Mountains Surrounding the Ashokan Reservoir," OFLA, undated document.

18. Jan Pleasants scrapbook, Robert Pleasants Collection, TOA; "Clearing and Grubbing the Ashokan Reservoir," *Catskill Water System News*, March 20, 1913, 169–70.

19. Board of Water Supply, *Annual Reports* (*Ninth Annual Report*, 1914), 124.

20. Steuding, *Last of the Handmade Dams*; Sickler, *Town of Olive through the Years, Part Two*, 20–21.

21. Board of Water Supply, *Annual Reports* (*Seventh Annual Report*, 1912), 103; Nimsgern, *Illustrated and Descriptive Account of the Main Dams and Dikes of the Ashokan Reservoir*, 14.

22. Board of Water Supply, *Catskill Water Supply*, 7. Blacks sometimes shared barracks with men of other ethnic backgrounds, but they always slept in their own rooms.

23. For a discussion of camp life, see a reprint of a report by the New York State Department of Labor in *Engineering News* 62 (August 5, 1909): 154–57.

24. See box JVS, folder John Watson Photograph Collection 1910, photo titled "Street Scene at Brown's Station," TOA.

25. Steuding, *Last of the Handmade Dams*, 44–50.

26. Shupe, Steins, and Pandit, *New York State Population, 1790–1980*, 215.

27. Provost, "Protection of New York's Water Supply from Pollution during Construction Work."

28. Board of Water Supply, *Catskill Water Supply*, 7.

29. "Works of Sanitation," song 38, from *Board of Water Supply Songs*, 1909.

30. Board of Water Supply, *Annual Reports* (*Eighth Annual Report*, 1913), 4–5.

31. Galusha, *Liquid Assets*, 151–52.

32. *Approximate Estimate of Cost of Delivering 500 Million Gallons of Water Daily into New York City from the Catskill Watershed*, box 103, folder Corresp. Oct. 1905, JRF, MIT Archives.

33. Board of Water Supply, *Annual Reports* (*Fourth Annual Report*, 1909), 4.

34. John Ripley Freeman to J. Waldo Smith, July 20, 1915, "Investigation Tour of Schoharie, Ashokan Reservoir, Northern Aqueduct, Hudson Crossing, Kensico, Park Hill, July 13, 14 and 15, 1915," box 104, folder NYC Board of Water Supply, Corr. 1915, JRF, MIT Archives.

35. Board of Water Supply, *Catskill Water Supply*, 5.

36. Ridgway, "Hudson River Crossing of the Catskill Aqueduct," 324.

37. Ridgway, *Robert Ridgway*, 166–75; Board of Water Supply, *Catskill Water Supply*, 36.

38. Contained in "The Holing Through of the Tunnel under the Hudson River," January 30, 1912, box 104, untitled folder (folder contains correspondence from 1911 and 1912), JRF, MIT Archives.

39. Hall, *Catskill Aqueduct*, 115–18. Long storage in reservoirs subject to high nutrient loading can reduce water quality. Several decades after the construction of Kensico Reservoir, New York City scientists encountered this problem there due to fecal deposition by geese.

40. Board of Water Supply, *Catskill Water Supply*, 31.

41. Troetel, "Suburban Transportation Redefined," 278, 289.

42. For Freeman's early views, see John Ripley Freeman to J. Waldo Smith, November 2, 1907, box 103, folder Corr. July–Sept. 1907, JRF, MIT Archives. His revised opinion is contained in John Ripley Freeman to Philip Sawyer, May 24, 1919, box 104, folder New York City Board of Water Supply, Corr. 1919–20, JRF, MIT Archives.

43. Fitzgerald, "Journey to Catskill Supply," 396.

44. Galusha, *Liquid Assets*, 114.

45. Board of Water Supply, *Annual Reports* (*Fourth Annual Report*, 1909), 6–7. See also White, *Catskill Water Supply of New York City*, 47–48.

46. Although the reservoir is still in place, it is used as part of the drainage system for underground tanks that lie beneath it.

47. A BWS song, "Ode to Bishop Falls," paid tribute to the disappearance of the local landmark. See *Board of Water Supply Songs*, Celebration of Beginning of Storage of Catskill Water, Song 13, JRF, MIT Archives.

48. In endorsing the city's proposal to divert water from the Schoharie Creek, John Freeman closely echoed Pinchot's emphasis on the need to take full advantage of nature's gifts and the utilitarian imperative to benefit the greatest number of people: "True conservation in its highest sense requires that every natural resource should be put to its best use," which he defined as "that of the City whose population composes one-half that of the entire State and which pays practically three-quarters of all the taxes which the State itself assesses." John Ripley Freeman Affidavit to State Conservation Commission, May 18, 1914, box 104 untitled folder (contains Freeman's correspondence for 1914), JRF, MIT Archives.

49. Cronon, *Changes in the Land*, chap. 1; Smith, *Virgin Land*, 189–96. In a similar vein, preservationists often downplayed the presence of Indians and others on land they hoped to turn into protected areas. See Spence, *Dispossessing the Wilderness*.

50. State Water Supply Commission, *Second Annual Report*, 81.

51. Hoffman, "Ulster County Bluestone," 122.

52. Ibid., 127.

53. Howe, *Old West Hurley Revisited*, 13, 39.

54. DuMond, *Walking through Yesterday*, 16.

55. Board of Water Supply, *City of New York Additional Water Supply*, 30–34. Edward Hagaman Hall argued that the greatest of Roman aqueducts "in technical difficulty, was, in comparison, like building houses with children's 'blocks.'" See Hall, *Catskill Aqueduct*, 5.

56. Longstreth, *Catskills*, 269. Vera Sickler, who retained childhood memories of the reservoir construction and later served as town historian of Olive—one of the two communities partially inundated by the Ashokan Reservoir—observed that Catskill residents did not share this perspective. "It is said the reservoir fitted in its place as if it were always there perhaps it was meant to be there in Gods [*sic*] almighty plan as some evidence was found of a pre glacial lake by the reservoir engineer surveyors. The people who lost their homes didn't think of it that way." See box GVS, notebook no. 26, Vera Sickler Papers, TOA.

57. The historical record is inconclusive regarding the existence of a Chief Ashokan. However, the literal meaning of Ashokan, "where waters converge," proved an apt name for the reservoir. See Pritchard, *Native New Yorkers*, 250–51.

58. Hall, *Catskill Aqueduct*, 115–18.

59. Raymond Carruthers, interview by Vera Sickler, October 13, 1977, box JVS, folder Handwritten Notes, Vera Sickler Papers, TOA.

60. Metz, *Lands Taken in Condemnation*, 27.

61. Stradling, *Making Mountains*, 94.

62. Ashokan Reservoir Awards, Martha Young Claimant, reel 118, 2626 (hereafter cited as ARA).

63. ARA, Clara Gricks Claimant, reel 117, 2022; David Stradling described the operations of the Bishop Falls house, a prominent boardinghouse submerged by the Ashokan Reservoir. See Stradling, *Making Mountains*, 94–95.

64. ARA, General Arguments, 3109.

65. Stradling, "Bishop Falls," 418–21. In urban settings, grocers' wives often ran boardinghouses, taking advantage of their access to food at wholesale prices. See Gamber, *Boardinghouse in Nineteenth-Century America*, 50–51.

66. Abbott, "Subway for Water." Quotations on 163.

67. See Davis, *Eden of the Catskills* (self-published pamphlet), OFLA.

68. Box JVS, folder Typed Notes, Vera Sickler Papers, TOA. The Ulster and Delaware Railroad described West Shokan as a "pleasant hamlet with churches, schools, stores, and many boarding houses where hundreds of city people pass the summer delightfully and at moderate cost." See Ulster and Delaware Railroad, *Catskill Mountains*, 12.

69. See Ulster and Delaware Railroad, *Catskill Mountains*, 10.

70. Sickler, *Town of Olive through the Years, Part 2*, 40.

71. A clear explanation of the process by which the city claimed and paid for land is included in Metz, *Lands Taken in Condemnation*, 13–21.

72. Rodgers, "In Search of Progressivism," 126–27; James Scott highlights the importance of legibility for the modern state. See Scott, *Seeing Like a State*, 2–3.

73. Board of Water Supply, *Annual Reports* (*Eighth Annual Report,* 1913), 22–23.

74. Ashokan Reservoir Awards (hereafter cited as ARA), Martha Young Claimant, reel 118, 2653.

75. ARA, Clara Gricks Claimant, reel 117, 2121.

76. ARA, Uriah Wood Claimant, reel 117, 1940–2013. Winston & Co. joined forces with another contractor, MacArthur Brothers, on the construction of the Ashokan Reservoir. Most residents referred to the contractors as "Winston."

77. ARA, Max Ferro Claimant, reel 185, 20, 199–200, 260.

78. ARA, Max Ferro Claimant, reel 185, 20, 247.

79. ARA, Thomas Donahoe Claimant, reel 41, 5596–604.

80. Board of Water Supply, *Annual Reports* (*Ninth Annual Report,* 1914), 124, 159.

81. Board of Water Supply, *Minutes of the Meetings*, 1920, 12, 28, 201; Board of Water Supply, *Minutes of the Meetings*, 1922, 138.

82. Board of Water Supply, *Annual Reports* (*Eighth Annual Report*, 1913), 21–23.

83. Shupe, Steins, and Pandit, *New York State Population, 1790–1980*, 113, 215.

84. DuMond, *Walking through Yesterday*, 147–48.

85. Carruthers, interview by Sickler. Vera Sickler Papers, TOA. Carruthers recalled a neighbor, Oliver MacAvoy, who refused to leave his home, and died the morning of his eviction day.

86. "Census Ashokan Reservoir," box X, folder Ashokan Reservoir Census 1913, TOA.

87. Quoted in Steuding, *Last of the Handmade Dams*, 101.

88. Ulster and Delaware Railroad, *Catskill Mountains*, 11–12; Catskill Enterprise, *Picturesque Catskill*.

89. Slichter, "Franklin D. Roosevelt's Farm Policy," 167–76.

90. "Brodhead's and Ashokan Reservoir," 104.

91. Bond, *Ulster County: Census Data*, 1–3.

92. Town of Olive Clerk's Office, *Step Back in Time*, 1997, videocassette.

93. "Removal of Bodies from the Ashokan Cemeteries," 89–90.

94. Steuding, *Last of the Handmade Dams*, 76; for information on Schoonmaker see "Census Ashokan Reservoir," box X, folder Ashokan Reservoir Census 1913, TOA.

95. Longstreth, *Catskills*, 270.

96. *Kingston Daily Freeman*, November 28, 1905, 1. Ironically, Clearwater was a major beneficiary of the commission process. He represented hundreds of Ulster County residents before commissions, earning substantial fees in the process.

97. Steuding, *Last of the Handmade Dams*, 89.

98. ARA, Martha Young Claimant, reel 118, 2604–89; ARA, Rebecca Bonesteel Claimant, reel 117, 2104–47.

99. ARA, Laura Every Claimant, reel 123, 7585.

100. Ibid., 7591.

101. Ibid., 7613.

102. Steuding, *Last of the Handmade Dams*, 92–93.

103. "Ashokan Awards Delayed," *Kingston Daily Freeman*, March 22, 1909, 5.

104. DuMond, *Walking through Yesterday*, 144.

105. "Predatory Taxation Threatens Ulster Co.," *Pine Hill Sentinel*, March 15, 1913, 2.

106. "Hearing on Tax Dodging Bill," *Kingston Daily Freeman*, January 24, 1913, 1.

107. Metz, *Lands Taken in Condemnation*, 22.

108. Carey, Traum, and Dupree, *Deep Water*, videocassette.

109. Sickler, *Town of Olive through the Years*, 77.

110. Board of Water Supply, *Report of the Board of Water Supply to the Mayor*, 9.

111. Diner, *Very Different Age*, 119–20.

112. Eisenach, *Lost Promise of Progressivism*, 3–5; Cohen, *Reconstruction of American Liberalism, 1865–1914*, 5–10.

113. Agricultural historian Deborah Fitzgerald describes a similar push by government officials and banks in the 1920s to get farmers to track their income and expenses, a critical step in the industrialization of agriculture. See Fitzgerald, "Accounting for Change."

114. On the systematization of American political and social life, see Hays, "Theoretical Implications," 19–21.

115. Van Valkenburgh and Olney, *Catskill Park*, 35–48.

116. By 2004, approximately 45 percent of the city's watershed lay within the Catskill Park. See Van Valkenburgh and Olney, *Catskill Park*, 87.

117. White, *Organic Machine*.

118. John Ripley Freeman to J. Waldo Smith, October 25, 1920, box 104, folder NYC Board of Water Supply, Correspondence, 1919–20, JRF, MIT Archives.

3. Droughts, Delays, and the Delaware

1. Henry Towne to John Freeman, February 13, 1919, box 104, folder NYC Board of Water Supply, Correspondence, 1919–20, JRF, MIT Archives.

2. A typical advertisement for an apartment rental boasted of a "large, comfortable, sunny, front room; running water; bath on same floor." *Brooklyn Daily Eagle*, March 12,

1920, 13. Important works on the physical transformation of the city in the early twentieth century include Hood, *722 Miles*, and Stern, Gilmartin, and Massengale, *New York 1900*.

3. Wurster, *Last Municipal Administration of Brooklyn*, 29.

4. Gibson, *Population of the 100 Largest Cities*, 13.

5. "Aqueduct Making Long Island Swamp," *New York Times*, February 23, 1917.

6. Department of Water Supply, Gas, and Electricity, *Some of the Things Accomplished by the Department, 1914–1917*, 6.

7. "Aqueduct Making Long Island Swamp."

8. Caro, *Power Broker*, 143–322.

9. Ibid., 157–59; Blakelock, *History of the Long Island State Parks*, 33.

10. Caro, *Power Broker*, 159.

11. Blakelock, *History of the Long Island State Parks*, 31.

12. Long Island State Park Commission, *Annual Report*, 1926, 40.

13. "Smith Dedicates Nassau Parkway," *New York Times*, November 7, 1927.

14. Long Island State Park Commission, *Annual Report*, 1925, 24.

15. For a brief overview of the connection between parks and water infrastructure, see John Mattera, "New York City's Water Parks," *Daily Plant*, September 9, 2004, http://nycgovparks.org/sub_newsroom/daily_plants/daily_plant_main.ph (accessed August 22, 2008).

16. For an analysis of the connection between public works and city planning, see Peterson, "Impact of Sanitary Reform," 83–103; and Schultz and McShane, "To Engineer the Metropolis." Philadelphia's City Hall is built on the grounds of a former pumping station, the city's major arts museum was a former reservoir, and much of Fairmount Park, the city's largest park, was acquired to reduce pollution entering the Schuylkill River, one of the city's two major water sources. On Philadelphia, see Mandarano, "Clean Water, Clean City."

17. Department of Parks of the City of New York, *Annual Report*, 1905, 130.

18. Department of Parks of the City of New York, *Annual Report*, 1916, 177; Department of Parks of the City of New York, *Annual Report*, 1927, 54.

19. "Nathan Straus, Park Protector—I," *New York Daily News*, June 24, 1930; Blackmar and Rosenzweig, *Park and the People*, 448–69.

20. Rogers, *Rebuilding Central Park*, 106–14.

21. "Holiday Excursions," *Harper's Weekly*, July 24, 1880, 471.

22. "High Bridge," *Harper's Weekly*, August 22, 1885, 555. For background on the reservoir and water tower, see *Altman Memorial Carillon High Bridge Park, New York City*, folder Highbridge, New York City Parks Department Library, Vertical Files (hereafter cited as NYCPDL).

23. "Highbridge Park," July 1999, folder Highbridge, NYCPDL.

24. Renner, *Washington Heights*, 94–98.

25. See Krieg, *Robert Moses*; Schwartz, *New York Approach*; and Ballon and Jackson, *Robert Moses and the Modern City*.

26. For an overview of Moses's tenure as parks commissioner in the 1930s, with a focus on the development of public swimming pools, see Gutman, "Equipping the Public Realm"; for Moses's use of "reassignment" see Ballon and Jackson, *Robert Moses and the Modern City*, 177.

27. On American public swimming pools in the twentieth century, see Wiltse, *Contested Waters*.

28. Moses snared one-seventh of all WPA spending for New York City. See Caro, *Power Broker*, 453.

29. Department of Parks of the City of New York, *Improvement and Expansion of New York City Park Facilities*, 2. Sanitation officials agreed that sewage, most of it originating within the city, was primarily responsible for the degraded condition of New York's harbor and rivers; see "Final Report of the Research and Engineering Committee" (Tri-State Treaty Commission), January 1932, in series A 1118–80, box 1, folder 3, New York State Archives (hereafter cited as NYSA), 6–20.

30. Goldman, *Building New York's Sewers*. New York City's unwillingness to spend significant sums treating wastes that would have no influence on the quality of the water it consumed was typical. Most cities opted to filter their water supply, which was much cheaper than maintaining the purity of supplies by aggressively treating wastewater. See Tarr, *Search for the Ultimate Sink*, 162–65.

31. Partnership for Parks, *The High Bridge and Highbridge Parks*, Highbridge folder, NYCPDL.

32. Gutman, "Equipping the Public Realm," 143.

33. Herbert Levenson, letter to Robert Moses, July 4, 1939, box 102464, folder 6 (Highbridge Park—Swimming), Department of Parks and Recreation Records, MACNY.

34. Jane Rush, letter to Mayor Fiorello La Guardia, August 12, 1938, box 102407, folder 61 (Highbridge Park), Department of Parks and Recreation Records, MACNY.

35. "Mr. Moses Keeps at It," *New York Times*, April 6, 1934.

36. Press release, City of New York Parks Department, September 11, 1937, Williamsbridge Oval folder, NYCPDL; "Promenade to Cap Huge Play Center," *New York Times*, May 9, 1935.

37. "A Plea for Cooperation among Playground Members," *Clarion*, February 1938, 2.

38. Quoted in Spann, *New Metropolis*, 3.

39. Page takes the title of his book from the economist Joseph Schumpeter's phrase "the creative destruction of capitalism." Page, *Creative Destruction*.

40. In his environmental history of Seattle, Matthew Klingle explains that the decision to tap rural watersheds altered not only the streams and rivers the city dammed, but the topography of Seattle itself. Equipped with a reliable supply of water delivered under high pressure, engineers used hydraulic power to level many of Seattle's hills, a task that would have been impossible using conventional methods. See Klingle, *Emerald City*, 86–104. On the importance of viewing cities in a regional context, see Mohl, "City and Region."

41. For a thoughtful treatment of the churning quality of modernism, see Berman, *All That Is Solid*.

42. For population data, see Gibson, *Population of the 100 Largest Cities*, 11–15; for information on surging consumption, see J. Waldo Smith to John Freeman, October 25, 1920, box 104, folder NYC Board of Water Supply, Correspondence, 1919–20, JRF, MIT Archives.

43. J. Waldo Smith to John Freeman, October 25, 1920.

44. Charles Newell Burch, *State of New Jersey, Plaintiff, v. State of New York and City of New York, Defendants*, 139–40 (hereafter cited as *Report of the Special Master, 1931*).

45. *Preliminary Report to the Board of Water Supply*, included in letter from Thaddeus Merriman to John Freeman, December 13, 1922, box 104, folder NYC Board of Water Supply, 1921–22, JRF, MIT Archives.

46. In fact, Merriman's high reputation in the engineering field rested both on his technical achievements, such as his experiments with Portland cement, and his leadership of the BWS. The dam forming the Rondout Reservoir was named the Merriman Dam in his honor.

47. *Preliminary Report to the Board of Water Supply*, 6.

48. Lane, *Argument of Merritt Lane in Opposition to the Tri-State Treaty*, 31.

49. The successful establishment of the Port Authority of New York and New Jersey in 1921 contrasts sharply with the inability to reach consensus regarding the Delaware. The uncertainty surrounding interstate water law and the discrepancy in timing help explain the divergent outcomes. New York and New Jersey both had a clear interest in promoting development of the metropolitan region and its transportation network. The future of the Delaware was murkier. If New York's legal bid to tap the stream failed, New Jersey and Pennsylvania would be in a better position than if they agreed to permit New York City to divert hundreds of millions of gallons a day. It was only after the city secured the legal right to divert Delaware water that the states established the Interstate Commission on the Delaware River Basin (INCODEL) to promote the orderly development of the Delaware basin region. On the establishment of the Port Authority, see Doig, *Empire on the Hudson*, 27–75.

50. Board of Water Supply to Mayor James Walker, contained in Board of Water Supply, *Report of the Board of Water Supply of the City of New York to the Board of Estimate and Apportionment*, 1.

51. Pugliese, "Citizen Group Activity in the Delaware River Basin," 155–58; Board of Water Supply, *Report of the Board of Water Supply of the City of New York to the Board of Estimate and Apportionment*, 4.

52. Board of Water Supply, *Opinion of Hon. John W. Davis*, 6–8.

53. Ibid., 10.

54. The prevailing water paradigm in western states was prior appropriation. Under prior appropriation, "beneficial uses carry a priority as of the date the use was initiated. In the event of a water shortage, the 'first-in-time' (the most senior water right) is 'first-in-right' (entitled to priority)." See Sherk, *Dividing the Waters*, 4.

55. Hufschmidt, *Supreme Court and Interstate Water Problems*, 21. For additional insight into contemporary perspectives on water law, see Interstate Commission on the Delaware River Basin, *INCODEL*, 47.

56. Hufschmidt, *Supreme Court and Interstate Water Problems*, 21–22.

57. *Report of the Special Master, 1931*, 1.

58. Philadelphia had long drawn on the Delaware for much of its water supply, but its intakes were located within the city. Opposing New York's bid for more Delaware water would prevent Philadelphia from tapping these same upstream sources in the future. It had long coveted these sources because they were much purer than downstream supplies.

59. *Report of the Special Master, 1931*, 30–31.

60. Ibid., 25–37.

61. *State of New Jersey, Complainant, the Delaware River Case*, 4:404.

62. *Report of the Special Master, 1931*, 68.

63. Ibid., 205.

64. Ibid., 159–76.

65. For Burch's conclusions, see ibid., 193–209. Reasonable use was the basis for determining equitable apportionment of a river's benefits, according to Burch.

66. Thaddeus Merriman to John Freeman, February 16, 1931, box 104, folder, BWS, Corr. 1931, JRF, MIT Archives.

67. Abel Wolman to Charles Burch, February 12, 1931, box 6.28, folder Delaware River Diversion Case, 1930–31, Ms. 105, Abel Wolman Papers, Special Collections, Milton S. Eisenhower Library, Johns Hopkins University (hereafter cited as WP).

68. Thaddeus Merriman to John Freeman, February 16, 1931.

69. For several examples, see *Report of the Special Master, 1931*, 25–26, 43–45, 65, 142.

70. Charles Burch to Abel Wolman, February 4, 1931, box 6.28, folder Delaware River Diversion Case, 1930–31, WP.

71. *Report of the Special Master, 1931*, 197.

72. Galusha, *Liquid Assets*, 213, 270.

73. H. G. Irwin, "Notes on Consumption," *Delaware Water Supply News*, June 15, 1939, 97–98; Committee on Additional Water Supply, *Report of Mayor O'Brien's Committee on Additional Water Supply*, 3.

74. "Water Conservation Program on a Vast Scale: New York City Fights a Water Shortage with Publicity, Inspections, Cooperation," *American City Magazine*, April 1940, 55–59.

75. H. G. Irwin, "Water Consumption: Sanitary Facilities," *Delaware Water Supply News*, August 15, 1939, 116.

76. La Guardia, "Ten Misconceptions," 651.

77. Works Progress Administration, *Survey of Water Consumption*.

78. Maurice P. Davidson, "The Logic of Universal Metering," *American City Magazine*, May 1935, 63.

79. *American City Magazine. A Report on the New York City Water Supply and the Effect Water Meters Would Have on That Supply*, 4.

80. Ibid.

81. Cole, "Water Waste and Its Detection."

82. "City Wastes Water," *New York Tribune*, January 16, 1910, 13.

83. Citizens Union, *Water Meters for New York City's Multiple Dwellings*, 2; "New Law for Water Tax Payments," *New York Times*, April 14, 1912; James G. Thurber, "Slogan 'Water Free as Air' Costs Millions for Standing as Specious Bar to Meters," *New York Evening Post*, February 17, 1927. Before working as a humorist and cartoonist for the *New Yorker*, Thurber worked as a newspaper reporter. The series he produced on water meters was one of his last major contributions to the *Post*.

84. Mayor William Gaynor to High Bridge Taxpayers' Alliance, reprinted in "Inspectors Find Many Water Leaks," *New York Times*, June 14, 1911.

85. *Annual Report of the Department of Water Supply, from Henry S. Thompson, Commissioner, to Mayor William Gaynor*, December 26, 1911, contained in Board of Aldermen, *Proceedings of the Board of Aldermen of the City of New York*, January 1 to March 19, 1912, 1:195–96.

86. James G. Thurber, "U.S. or State May Order Metering," *New York Evening Post*, February 15, 1927.

87. *State of New Jersey, Complainant, the Delaware River Case*, 1:8.

88. *Report of the Special Master, 1931*, 44. Burch also rejected New Jersey's argument for universal metering on the grounds that it was up to New York City to decide this on its own. See 42–43.

89. Committee on Additional Water Supply, *Report of Mayor O'Brien's Committee*, 9.

90. Waring, "Water Famine in Brooklyn," 134.

91. Thurber, "U.S. or State May Order Metering," 4.

92. "Water Conservation Program on a Vast Scale: New York City Fights a Water Shortage with Publicity, Inspections, Cooperation," *American City Magazine*, April 1940, 55–59.

93. "Water of the City Guarded by a Ditty," *New York Times*, January 17, 1940.

94. Goldberg's cartoons originally appeared in the *New York Sun*, "Water Shortage Suggestions," February 19, 1940.

95. Mayor's Executive Committee on Administration, *Report to the Mayor Concerning the Water Supply and Immediate Need for Conservation*, 2.

96. Division of Analysis, New York Bureau of the Budget, *Study of Water Conservation—Metering as a Basis for Charges for the Use of Water*, 14.

97. Gill, "Gathered Waters," 35.

98. "Reform in Air-Conditioning," *New York Herald Tribune*, March 25, 1950. The regulations required air conditioners to be of the circulating variety, that is, to reuse water for cooling rather than draw on additional supplies.

99. Gill, "Gathered Waters," 35.

100. Susanne Bedell, letter to the editor, *New York Herald Tribune*, March 22, 1950.

101. "Reform in Air-Conditioning."

102. Greater New York Taxpayers Association to Assistant Mayor William Reid, February 15, 1950, box 173, folder 1879, Mayor William O'Dwyer Subject Files, MACNY.

103. Press Release from Mayor William O'Dwyer, February 14, 1950, box 173, folder 1879, Mayor William O'Dwyer Subject Files, MACNY.

104. Mayor William O'Dwyer to Irving Huie, January 11, 1950, box 173, folder 1879, Mayor William O'Dwyer Subject Files, MACNY.

105. *Magazine of the New York Philanthropic League*, March 1950, 2, in box 174, folder 1880, Mayor William O'Dwyer Subject Files, MACNY.

106. In December 1949, a Detroit man volunteered his rainmaking services to New York. City officials spurned his offer. What had seemed absurd in December became a critical component of New York's water supply strategy in March. See Gill, "Gathered Waters," 34.

107. Richard K. Winslow, "Clouds Elude Rain-Maker Plane after Fog Delays Take-Off," *New York Herald Tribune*, March 29, 1950.

108. A. Finn Dickinson to Mayor William O'Dwyer, April 14, 1950, box 174, folder 1881, Mayor William O'Dwyer Subject Files, MACNY.

109. "City Now Skeptic on Rain-Making," *New York Times*, November 5, 1951. Howell claimed that he had boosted precipitation on the city's watersheds by 14 percent in the latter part of 1950.

110. Several observers of the city's water strategy identified the division of responsibility between two agencies as a principal cause of New York's lackluster commitment to conservation. See, for example, Perrin, "New York Drowns Another Valley," 83.

111. As one report observed, conservation appeals "are effective only for short periods and lose their effectiveness when repeated often." See Citizens Union, *Water Meters for New York City's Multiple Dwellings*, 4.

112. Gibson, *Population of the 100 Largest Cities*, 13–15.

4. Back to the Supreme Court

1. Crane, *Transformation of the Avant-Garde*; Bender, *New York Intellect*, 321–40.

2. On Wagner's role in expanding public housing, see Bloom, *Public Housing That Worked*, 123–27.

3. Robert Moses expressed his support for the board's policy of tapping mountain-fed supplies, observing that "our drinking water, like Caesar's wife, should be above suspicion." "Ripples of Doubt Cast on City's Water Ideas," *New York World-Telegram and the Sun*, April 30, 1952.

4. Gibson, *Population of the 100 Largest Cities*, 11–15.

5. For a comprehensive contemporary overview of Westchester's water supply, see Carman, *Water Supply Problems*.

6. Board of Water Supply, *1,820,000,000 Gallons per Day*, 18.

7. Jacobs, *Death and Life of Great American Cities*. Historians have begun to take note of the significant overlap between the rise of ecology and Jacob's perspective on urban life. See Kinkela, "Ecological Landscapes of Jane Jacobs and Rachel Carson."

8. On the Lower Manhattan Expressway, see Ballon and Jackson, *Robert Moses and the Modern City*, 212–14.

9. The literature on the decline of cities in twentieth-century America is vast. See Sugrue, *Origins of the Urban Crisis*, and Rae, *City*. On New York, see Caro, *Power Broker*; Freeman, *Working-Class New York*; and Cannato, *Ungovernable City*.

10. C. R. Cox to Abel Wolman, September 6, 1950, box 6.29, folder 1950 NYC Mayor's Committee, Water Supply Correspondence, WP.

11. BWS president Irving Huie highlighted the long-term commitment to tapping upland sources: "These expert opinions have determined the policy of the City since the inception of the Board of Water Supply." Irving Huie to Abel Wolman, June 6, 1951, box 6.29, folder 1951 NYC Mayor's Committee . . . Water Supply Correspondence, WP.

12. Board of Water Supply petition to New York State Water Power and Control Commission, quoted in Norman Kenney to Engineering Panel on Water Supply, September 5, 1950, box 6.28, untitled folder, WP.

13. Harold Riegelman, counsel to Citizens Budget Commission, to Board of Water Supply commissioner Irving Huie, February 15, 1950, Vertical Files, Loeb Library, Harvard University.

14. Harold Riegelman, counsel to the Citizens Budget Commission, to Water Power and Control Commission, October 4, 1950, Vertical Files, Loeb Library, Harvard University.

15. Riegelman to Huie, February 15, 1950.

16. Henry Z. Pratt Jr., division engineer, "Compilation of Studies Report to December 1, 1949," box 6.28, folder New York City Water Supply Abel Wolman, WP.

17. Riegelman to Huie, February 15, 1950.

18. A detailed account of the city's arguments before the commission is contained in Corporation Counsel of the City of New York, *In the Matter of the Application of the City of New York to the Water Power and Control Commission*, in box 6.28, folder 1950 NYC Mayor's Committee . . . Water Supply Correspondence, WP.

19. Riegelman to Water Power and Control Commission, October 4, 1950.

20. "The City's Water Future," *New York Times*, November 20, 1950.

21. Frank Marston, *Report to Roger W. Armstrong, Consulting Engineer Board of Water Supply, City of New York upon Quality of Hudson River Water Opposite Shaft 6 and Its Use as an Emergency Water Supply for New York City*, November 8, 1948, box 6.29, folder 1950 NYC Mayor's Committee . . . Water Supply Correspondence, WP.

22. Corporation Counsel of the City of New York, *In the Matter of the Application of the City of New York*, 3.

23. Mayor's Committee on Management Survey, *Future Water Sources*, xiv.

24. Thorndike Saville to L.R. Howson, consulting engineer, June 1, 1951, box 6.29, folder 1951 NYC Mayor's Committee . . . Water Supply Correspondence, WP.

25. Mayor's Committee on Management Survey, *Future Water Sources*, 79.

26. Ibid., 60.

27. Louis Howson to Thorndike Saville, December 3, 1951, box 6.29, folder 1951 NYC Mayor's Committee . . . Water Supply Correspondence, WP. "Wastes" in this context refers to water spilling over the dam when reservoirs are filled to capacity due to high rainfall.

28. Mayor's Committee on Management Survey, *Future Water Sources*, 31, 71.

29. Thorndike Saville to members of the Engineering Panel, November 27, 1951, box 6.29, folder 1951 NYC Mayor's Committee . . . Water Supply Correspondence, WP.

30. "Experts Boiling as City Puts Lid on Water Study," *New York World-Telegram and the Sun*, November 30, 1951.

31. Irving Huie expressed the standard formulation when he observed that tapping the Hudson would mean abandoning "substantially pure upland sources" for "heavily polluted and contaminated water from the Hudson." "Hudson Water Plan Rapped," *New York Daily News*, May 2, 1952.

32. Engineering Panel on Water Supply of the Mayor's Committee on Management Supply to Bernard Gimbel, March 7, 1952, *Comments on Board of Water Supply Report of October 30, 1951*, box 6.29, folder New York City Water Supply Mayor's Committee 1950, '51, '52, '53, WP.

33. "Ripples of Doubt Cast on City's Water Ideas," *New York World-Telegram and the Sun*, April 30, 1952.

34. Robert Moses to Bernard Gimbel, December 19, 1951, Mayor Vincent Impelliterri Subject Files, MN 45313, box 25, roll 13, folder 246 (Delaware River Water Supply Project), MACNY.

35. "Water Supply for the Future," *New York Herald Tribune*, March 12, 1955.

36. "Psychological palatability" comes from Noel Perrin. Noel Perrin to Thorndike Saville, September 27, 1962, box 6.29, folder 1956–59 NYC Water Supply, WP.

37. New York Department of Water Supply, Gas and Electricity, *Analysis of Report of Engineering Panel on Water Supply*, 30.

38. The panel listed eight reasons for preferring the Hudson over Cannonsville. Only one of these—the eighth and final item on the list—mentioned the impact on the Catskills. Mayor's Committee on Management Survey, *Future Water Sources*, xiv.

39. Perrin, "New York Drowns Another Valley," 76.

40. On the INCODEL Plan, see Malcolm Pirnie Engineers, *Report on the Utilization of the Waters of the Delaware River Basin*. On the collapse of INCODEL proposal, see Martin, *Metropolis in Transition*, 118–19; and "High Court to Get State Water Plea," *New York Times*, October 30, 1951.

41. Kurt F. Pantzer, *State of New Jersey, Complainant, vs. State of New York, and City of New York, Defendants, Commonwealth of Pennsylvania, and State of Delaware, Intervenors. Report of the Special Master Recommending Amended Decree*, 26 (hereafter cited as Pantzer Report).

42. Ibid., 37–39.

43. Ibid., 22; Hufschmidt, *Supreme Court and Interstate Water Problems*, 56.

44. *State of New Jersey, Complainant, against State of New York and City of New York*, 3:1061–72.

45. City Planning Commission Member Goodhue Livingston Jr. expressed this sentiment quite clearly: "The taxpayers of the city are probably unaware that a Cannonsville to be built exclusively with their money will save the taxpayers of New Jersey millions of dollars. However, New Jersey is aware of this as can be ascertained if the statements recorded in the first Special Session of the New Jersey Senate No. 12 December 4th, 1953 are read." Dissenting statement of Goodhue Livingston Jr., box 6.29, folder 1954 NYC Water Supply, WP.

46. William Schnader to Abel Wolman, December 29, 1953, box 6.29, folder 1952–53 Mayor's Committee . . . Water Supply Correspondence, WP.

47. Pantzer Report, 107.

48. *State of New Jersey, Complainant, against State of New York and City of New York*, 1:225.

49. Ibid., 1:233.

50. Board of Water Supply Transcripts Relating to the Pepacton and Cannonsville Reservoirs, Paul Cooper Jr. Archives, Stevens-German Library, Hartwick College, Oneonta, NY, Claim of George McMurray, vol. 111, 15051–78 (hereafter cited as BWS Transcripts).

51. BWS Transcripts, Claim of Charles and Florence Webb, vol. 253, 7528–37.

52. Francis, *Land of Little Rivers*; Van Put, *Trout Fishing in the Catskills*.

53. In 1950, the state passed legislation to protect two of these streams, the Willowemoc and the Beaverkill. See "Bill Passed to Save Two Trout Streams," *Delaware Republican Express*, April 13, 1950.

54. BWS Transcripts, Claim of Merritt Stuart, vol. 14, 9471.

55. Ibid., 9471–72.

56. New York State Senate, Committee on the Affairs of the City of New York, *Report on Legislation to Revise Condemnation Procedures*, 44.

57. BWS Transcripts, Claim of Merritt Stuart, vol. 14, 9402–15.

58. Ibid., 9515.

59. Ibid., 9409.

60. Ibid., 9942.

61. BWS Transcripts, Claim of Howard and Frances Dingee, vol. 262, 1642–44.

62. BWS Transcripts, Claim of Merritt Stuart, vol. 14, 9448.

63. Ibid., 9537.

64. New York State Department of Environmental Conservation, Bureau of Fisheries, *Fishery Resources of the Upper Delaware Basin*, I-9.

65. Ibid., I-10.

66. Westchester County Water Agency, *1976 Annual Report*, 1. Westchester was the only county where the city's system became the primary source of water supply. Therefore, I focus on it rather than other counties where the role of the city's supply was peripheral.

67. Temporary State Commission on the Water Supply Needs of Southeastern New York, *Alternative Futures: A Re-Evaluation*, 8.

68. William Meadowcraft, *Preliminary Report on Water Supply, Report No. 2*, October 30, 1934, Commission on Government Records, 1921–56, series 164, A-0210 (50) L, folder "Reports, Water," Carl Pforzheimer Papers, Westchester County Archives (hereafter cited as WCA).

69. "To Ask Legislature to Aid Water Plans," *New York Times*, May 25, 1911.

70. Franko, *Your Mount Vernon Board of Water Supply*, 127–28.

71. For an excellent analysis of the long-term process of suburbanization, see Hayden, "What Is Suburbia?"

72. Meadowcraft, *Preliminary Report*.

73. Ibid.

74. Franko, *Your Mount Vernon Board of Water Supply*, 58.

75. Philip Anderson, *Daily Argus*, September 4, 1947, quoted in Franko, *Your Mount Vernon Board of Water Supply*, 50.

76. Carman, *Water Supply Problems*, 4.

77. Ibid., 9.

78. Canning, "Westchester County since World War II; Carman, *Water Supply Problems*, 4; Westchester County Department of Planning, *Databook: Westchester County, New York*, 3.

79. Westchester Water Works Conference, Water Resources Committee, *Report on the Water Resources of Westchester County, New York*, 7, Miscellaneous Files, series G-30, folder G 30–75, WCA.

80. Meadowcraft, *Preliminary Water Report*.

81. *Westchester County Environmental Health Services Annual Report, 1961*, series 199, Environmental Health Services Annual Reports, A-0272 (7) F, folder 1961 Environmental Health Services Annual Report, WCA.

82. *Westchester County Environmental Health Services Annual Report, 1973*, series 199, Environmental Health Services Annual Reports, A-0272 (8) F, folder 1973 Environmental Health Services Annual Report, WCA.

83. Westchester County Water Agency, *1976 Annual Report*, 3.

84. Hudson River Museum, *Old Croton Aqueduct*. The park's website is http://nysparks. state.ny.us/parks/96/details.aspx.

85. McMahon, "Westchester from the Roaring Twenties to V-J Day."

86. U.S. Senate Committee on Interior and Insular Affairs, *Northeast Water Crisis: Hearing before the Committee on Interior and Insular Affairs*, Eighty-Ninth Congress, first session, September 8, 1965, 3.

87. Groopman, "Effect of the Northeast Water Crisis," 37–47.

88. Ibid.

89. Hogarty, *Delaware River Drought Emergency*, 3.

90. For background on the DRBC, whose roots could be traced in part to the flood of 1955, see Herbert Howlett, "The Delaware River Basin Commission: An Experiment in Government," speech delivered at the Chesapeake Section of the American Water Works Association Meeting, September 10, 1964, contained in vol. 1, tab 3, Herbert Howlett Papers, Special Collections and University Archives, Rutgers University Libraries.

91. Martin, *Metropolis in Transition*, 121–24. The states formed the DRBC in part to stave off federal water pollution standards. On the federal role, see Andrews, *Managing the Environment*, 204–6.

92. Martin, *Metropolis in Transition*, 124.

93. Between 1950 and 1960, overall consumption increased by almost 150 MGD despite a loss of almost one hundred thousand residents. For a historical overview of New York City's water consumption, see Judd, *How Much Is Enough?* 26–27.

94. Hogarty, *Delaware River Drought Emergency*, 13–15.

95. Homer Bigart, "Mayor Inspects City Reservoirs," *New York Times*, July 14, 1965.

96. Senate Committee on Interior and Insular Affairs, *Northeast Water Crisis*, 3–4.

97. Herbert Howlett, "The Drought: Some of Its Lessons," speech delivered at Annual Conference of Water Resources Association / DRBC, November 8–9, 1965, vol. 2, doc. 1, Herbert Howlett Papers, Special Collections and University Archives, Rutgers University Libraries.

98. Adolph Katz, "Goddard Charges New York City with Pirating Water from the Delaware," *Philadelphia Evening Bulletin*, June 27, 1965.

99. Albert, *Damming the Delaware*, 14, 37.

100. John M. McCullough, "'Squeeze Play' Hinted in N.Y. Diversion Plan," *Philadelphia Inquirer*, December 4, 1952.

101. Hogarty, *Delaware River Drought Emergency*, 36.

102. "New York's Water Piracy," *Philadelphia Evening Bulletin*, July 3, 1965.

103. Water Resources Council, *Report to the President: A Reappraisal of Drought in Northeastern United States*, September 7, 1965, contained in Senate Committee on Interior and Insular Affairs, *Northeast Water Crisis*, 78.

104. Hogarty, *Delaware River Drought Emergency*, 33.

105. "The People-Water Crisis," *Newsweek*, August 23, 1965, 50.

106. "Excerpts from Wagner Talk on Water," *New York Times*, August 14, 1965; "Cloud Seeder Offers Water at Penny for 1,000 Gallons," *New York Times*, August 14, 1965.

107. Water Resources Council, *Report to the President*, 78.

108. Hogarty, *Delaware River Drought Emergency*, 44.

109. Ibid., 47–48.

110. Ibid., 35.

111. Lyndon Johnson, *Remarks of the President before the Water Emergency Conference*, August 11, 1965, contained in Senate Committee on Interior and Insular Affairs, *Northeast Water Crisis*, 90.

112. Mayor Robert Wagner press release, "Robert Wagner's Remarks to President's Special Task Force," Mayor Robert Wagner Subject Files, box 309, folder 3640, Water Supply (2) 1965, MACNY.

113. "Water: No Magic Remedy," *New York Times*, August 19, 1965.

114. John Berger, letter to the editor, *New York Times*, August 3, 1965.

115. When metering was finally implemented in the 1990s, landlords did take significant steps to curb water waste. See chapter 6. On the impracticality of metering individual apartments, see John Geyer to Engineering Panel, June 23, 1950, box 6.28, folder New York City Water Supply Abel Wolman, WP. The DWS supported metering, but it lacked the clout required to get top municipal officials to implement metering. On the department's support of metering multiple-unit dwellings, see "Plug Those Leaks!" *New York World-Telegram and Sun*, May 15, 1952.

116. Mayor's Panel on Water Supply, *Universal Metering for New York City*, Mayor Robert Wagner Subject Files, box 309, folder 3639, Water Supply (1), 1965, MACNY.

117. For the preliminary findings of the panel, see press release of September 2, 1965, Mayor Robert Wagner Subject Files, box 309, folder 3639, Water Supply (1), 1965, MACNY.

118. Mayor Robert Wagner press release, September 12, 1965, Mayor Robert Wagner Subject Files, box 309, folder 3639, Water Supply (1), 1965, MACNY.

119. Hirshleifer, De Haven, and Milliman, *Water Supply: Economics, Technology, and Policy*, 262.

120. See, for example, Gottehrer, *New York City in Crisis*.

121. Greater Philadelphia Movement, *A Position Paper on Delaware River Water*, 17.

122. David Grann, "City of Water," *New Yorker*, September 1, 2003, 88–109.

123. Water Power and Control Commission, *Decision of the Water Power and Control Commission on New York City's Application*, Interstate Basin Commission Administration Files, series A1118, box 1, folder B8, NYSA.

124. Scott, *Seeing Like a State*, 20, 46.

125. Completed just as environmentalism was beginning to flower, the Cannonsville Reservoir was conceived in a decidedly pre-ecological era. In 1969, two years after Cannonsville water began to flow into New York City, Congress passed the National Environmental Policy Act, NEPA, which required the federal government to perform environmental impact statements documenting the likely effects of government construction and other projects. By factoring ecological consequences into the democratic equation, NEPA and similar state legislation revolutionized government accountability, empowering environmentalists and community activists. See Flippen, *Nixon and the Environment*, 226, 48–53; Melnick, "Risky Business."

126. Robert Moses clearly articulated the logic behind bond funding: "I am for continuing to issue bonds exempt from the debt limit to provide a permanent gravity system from Cannonsville, as opposed to begging every year for expense budget items to maintain and operate gadgets to purify polluted Hudson water." Robert Moses to Mayor Impellitteri, April 25, 1952, Mayor Vincent Impelliterri Subject Files, MN 45352, box 104, roll 52, folder 1271 (Water Shortage 1952), MACNY.

127. A typically skeptical editorial noted, "The Board of Water Supply, which might find the need for its continuance in lifetime jobs removed by any permanent solution of the water problem, takes violent issue with the engineers." "New York's Water Supply," *New York World-Telegram and the Sun*, December 11, 1951.

128. The board embraced what James Scott has called a "high-modernist ideology," a supreme faith in scientific progress, the satisfaction of human needs, and control of nature. As Scott observes, believers in high modernism "tended to see rational order in remarkably visual aesthetic terms." A new Delaware reservoir fit the bill; pumping and then cleaning Hudson water simply did not square with the board's vision of beautiful and rational planning. See Scott, *Seeing Like a State*, 4.

129. William H. Rudy, "Row over Water Supply Nears Boiling Point," *New York World-Telegram and the Sun*, December 10, 1951.

5. The Water System and the Urban Crisis

1. Rogers, *Rebuilding Central Park*, 114. On Elton John, see "Crowd of 400,000 Sets Record for Park," *New York Times*, September 14, 1980; on Simon and Garfunkel, see Paul L. Montgomery, "Simon-Garfunkel Reunion Jams Central Park," *New York Times*, September 20, 1981.

2. Van Put, *Trout Fishing in the Catskills*, 387.

3. Capossela, *Good Fishing in the Catskills*, 9.

4. On the many problems confronting New York in the mid-1960s, see Gottehrer, *New York City in Crisis*.

5. For an analysis of the fiscal crisis and its lasting effects, see Freeman, *Working Class New York*, 256–90.

6. The Catskill village of Fleischmanns once boasted more than fifty hotels, but by the mid-1970s only four remained. See Mitchell, *Catskills*, 83–84. New York State Comptroller Arthur Leavitt noted this trend in the early 1970s: "There is a remarkable boom in farm real estate, at least when reachable by the jaded denizens of our metropolitan areas." Leavitt speech to Association of Towns, Record Group 15, series 10.3, Counsel's Office—Robert R. Douglass, box 54, folder 612, Nelson A. Rockefeller Gubernatorial Papers, Rockefeller Archive Center.

7. "Water Quality Protection," *Catskill Center News*, Summer 1988, 14.

8. For the classic contemporary expression of ecological ideas, see Commoner, *Closing Circle*.

9. Rockefeller championed cleaning up the state's polluted waterways and scored a major victory in 1965, when voters overwhelmingly approved a $1 billion bond measure for his Pure Waters program to upgrade sewage treatment plants.

10. For a good overview of the rise of environmental regulation, see Andrews, *Managing the Environment*, chap. 12.

11. On the central role of technology in environmental regulation, see Milazzo, *Unlikely Environmentalists*.

12. The clearest explanation of the demand of middle-class Americans for improved environmental amenities and a focus on quality over quantity is Hays, *Beauty, Health, and Permanence*.

13. Andrews, *Managing the Environment*, 202. On the growth of the environmental movement, see Shabecoff, *Fierce, Green Fire*.

14. Statement by Joe Kilpatrick at public hearing held in Oneonta, NY, March 9, 1977, series Lo131–78, Hearing Files and Transcripts, 1975–77, Legislative Assembly, Standing Committee on Environmental Conservation, box 1, folder 5, NYSA.

15. New York City's population decreased by more than eight hundred thousand people in the 1970s. See Gibson, *Population of the 100 Largest Cities*, 11–15.

16. Given this context, Saul Steinberg's famous *New Yorker* cover depicting New York City as the center of the world (with China, Japan, and Russia hovering discreetly in the background) seems particularly ironic. Even New Yorkers were no longer convinced of their centrality. See *New Yorker*, March 29, 1976.

17. Peter Grant, "Development Surge Upstate a Concern for City's Water," *New York Observer*, November 16, 1987.

18. "New York City Reviews Its Watershed, With Eye to Selling Surplus Land," *New York Times*, April 4, 1977.

19. Despite the commitment to construct a filtration plant for Croton waters, a lack of financing and disputes over site selection severely delayed construction of the plant. It is currently slated for completion in 2013.

20. On development in the Croton watershed, see Goldstein and Marx, *Under Attack*.

21. On the push to abolish the board, see Charles Kaiser, "Goldin Calls for End to City Water Board as an 'Anachronism,'" *New York Times*, December 19, 1977. The Department of Environmental Protection was created by the division of a "superagency," the Environmental Protection Administration, into two separate agencies: DEP and the Sanitation Department. Operating the city's water system remains DEP's most high-profile responsibility.

22. Mayor Ed Koch to DEP Commissioner Francis McArdle, July 13, 1978, Mayor Ed Koch Departmental Correspondence, box 141, folder 1 (Environmental Protection), MACNY.

23. DEP Commissioner Francis McArdle to Mayor Ed Koch, August 10, 1978, Mayor Ed Koch Departmental Correspondence, box 141, folder 1 (Environmental Protection), MACNY.

24. Francis McArdle to Mayor Ed Koch, Drought Watch Update (undated, but probably November 1980), Mayor Ed Koch Departmental Correspondence, box 141, folder 4, 1980, MACNY.

25. DEP Commissioner Joseph McGough Jr. to Mayor Ed Koch, March 5, 1985, Mayor Ed Koch Departmental Correspondence, box 142, folder 8 (Environmental Protection 1985), MACNY.

26. Albert F. Appleton, "Post-Delaware Era," 221.

27. Peter Borrelli, discussion with the author, February 2006. On the growing appeal of the Catskills to city residents, see Grant, "Development Surge."

28. Vincent Coluccio, discussion with the author, June 2008. In the mid-1980s, there were over seventy sewage treatment plants in the city's watersheds. As development increased, additional plants were constructed.

29. Ibid.

30. The state Department of Environmental Conservation had the primary responsibility for overseeing the operation of sewage treatment plants.

31. Department of Environmental Protection Commissioner Harvey Schultz to Mayor Ed Koch, September 16, 1988, Mayor Ed Koch Departmental Correspondence, box 144, folder 1, MACNY.

32. Coluccio, discussion.

33. Coluccio, discussion; Gordon and Kennedy, *Legend of City Water*, 27.

34. For a brief history of the DEC, see www.dec.ny.gov/about/9677.html; see also Birkland et al., "Environmental Policy in New York," 396–400. Schultz to Koch, September 16, 1988.

35. "Background Information Related to NYS DEC Adjudicatory Hearing on Delhi Sewage Treatment Plant SPDES Permit," box 6.30, folder 1986 NYC Delhi Case, WP.

36. Mark McIntyre, "Pure Water Remains His Passion," *Newsday*, December 30, 1986.

37. Abel Wolman, "The New York City Water Supply System," October 12, 1986, box 6.30, folder 1987 NYC Delhi Case, WP. Articles about the system invariably began by highlighting the traditionally high quality of the city's water. A 1987 article typified this discourse: "New York City's water supply, which consistently wins national and statewide competitions for quality . . ." See Grant, "Development Surge."

38. McIntyre, "Pure Water Remains His Passion."

39. Vincent Coluccio to Abel Wolman, January 9, 1987, box 6.30, folder 1987 NYC Delhi Case, WP.

40. Albert F. Appleton, "Post-Delaware Era," 220.

41. Perlee, *Future Cost of Water in New York City*, 1.

42. Mark McIntyre, "New York City Drinking Water, Trouble on Tap," *Newsday*, August 21, 1988.

43. Ibid.

44. The EPA did not publish the federal regulations detailing the water quality criteria cities would have to meet to avoid filtration until 1989. The Surface Water Treatment Rule requiring filtration or enhanced watershed protection was issued to implement the amendments to the Safe Drinking Water Act passed by Congress in 1986. For details on the SWTR and a subsequent set of amendments to the law, see National Research Council, *Watershed Management for Potable Water Supply*, 107–15.

45. Harvey Schultz to Deputy Mayor Robert Esnard, April 7, 1988, Mayor Ed Koch Departmental Correspondence, roll 61, box 144, folder 1 (Environmental Protection), MACNY.

46. Coluccio, discussion.

47. McIntyre, "New York City Drinking Water."

48. Environmental lawyer Robert F. Kennedy Jr. called on the city to improve enforcement of existing regulations, devise a new set of regulations relevant to contemporary development threats (the old ones, he observed, "seemed straight out of Andy of Mayberry"), and purchase more lands in its watersheds to protect water quality. See Robert F. Kennedy Jr., "New York City's Water: Down the Drain," *New York Times*, August 22, 1989.

49. McIntyre, "New York City Drinking Water," 34.

50. Eric Greenfield, interview by Nancy Burnett, *BTS*, November 7, 1995, 67.

51. "Olive Assessors Force New York City to Withdraw Demand for $1,000,000 Ashokan Slice," *Leader*, October 4, 1946, Town of Olive Records, Roll 65, NYSA.

52. Perry Shelton, interview by Nancy Burnett, *BTS*, February 20, 1996, 161–62.

53. Tobie Geertsema, untitled article, *Woodstock Times*, July 11, 1976.

54. Daniel Logan, "The Ashokan: A Winter View," *Catskill Center News*, Winter 1989, 9.

55. R. J. Kelly, "Side by Side: The Moral of Traver Hollow: Two Bridges Are Better Than One," *Woodstock Times*, November 13, 1980.

56. Shelton, interview, 157.

57. Cecil E. Heacox, "Liquid Assets," *New York State Conservationist*, June–July 1950, 2–9.

58. Hanover, *Guide to Trout Streams in the Catskill Mountains*, 6.

59. Cronon, *Nature's Metropolis*. Cronon offered a wonderfully detailed analysis of the wide-ranging environmental repercussions of Chicago's development. For other studies that connect urban development to changes in the countryside, see Klingle, *Emerald City*; Logan, *Desert Cities*.

60. The human tendency to simplify nature, both deliberately and unwittingly, is a favorite theme of environmental historians. See John McPhee's description of flood control on the Mississippi River. McPhee, *Control of Nature*.

61. For background on the Catskill Forest Preserve, see Van Valkenburgh and Olney, *Catskill Park.*

62. Norm Van Valkenburgh, discussion with the author, April 2008.

63. Ibid.

64. Van Put, *Trout Fishing in the Catskills,* 387; Capossela, *Good Fishing in the Catskills,* 6.

65. After their construction, reservoirs often take a few years to break down organic matter that accumulates in their basin. See McIntyre, "New York City Drinking Water"; Bureau of Fisheries, New York State Department of Environmental Conservation, "Fishery Resources of the Upper Delaware Basin," January 1975, series 16841–92, Division Director's Subject and Correspondence Files, 1962–96, Division of Fish, Wildlife, and Marine Resources, box 10, folder 8, NYSA.

66. "Hudson River OK for Water Supply," *Delaware Republican Express,* April 27, 1950.

67. "Hearing Opens on West Branch Dam," *Delaware Republican Express,* March 23, 1950.

68. "The Delaware River Basin," *Delaware Republican Express,* March 30, 1950.

69. The "blue line" defined the borders of the Catskill Park, a patchwork of public and private holdings. All state-owned land in the park is part of the Catskill Forest Preserve and is off-limits to logging and most development. The state confined its land acquisition efforts to the area within the blue line.

70. McIntyre, "New York City Drinking Water"; Van Valkenburgh, discussion.

71. "Draft: Water Quality Guidance West Branch Delaware Model Implementation Program," May 18, 1978, New York State Department of Environmental Conservation, series 16841–92, box 9, folder 17, NYSA.

72. Schultz to Koch, September 16, 1988.

73. "Meeting on New York City Reservoirs in the Upper Delaware River Basin," December 1974, series A 1120–80, New York State Department of Environmental Conservation, Drainage Basin Subject Files, box 4, folder 1, NYSA.

74. Capossela, *Good Fishing in the Catskills,* 45.

75. Buerle, *Water Resource Issues of the Catskill Region,* 58; Van Put, *Trout Fishing in the Catskills,* 388.

76. "Draft Fisheries Management Plan for the Upper Delaware River Tailwaters," New York State Department of Environmental Conservation, series 16841–92, box 10, folder 8, NYSA.

77. J. Douglas Sheppard, draft of article for *Conservationist,* May 4, 1978, series 16841–92, box 13, folder 9, NYSA.

78. Buerle, *Water Resource Issues,* 51.

79. Jim Serio, discussion with the author, April 16, 2008.

80. Borrelli, *Catskill Center Plan,* 49.

81. Heacox, "Liquid Assets," 8; Hanover, *Guide to Trout Streams,* 10.

82. Francis, *Land of Little Rivers,* 93.

83. O'Neil, "In Praise of Trout," 105.

84. Ibid., 106.

85. Chuck Schwartz, president, Phoenicia Fish and Game Association, to Fred Faerber, president, Ulster County Federated Sportsmen Clubs, November 1, 1974, series L0131–78, Legislative Assembly, Standing Committee on Environmental Conservation, box 1, folder 8 (Esopus Creek Fishkill), NYSA. New York City challenged reports of fish kills because fishermen were generally unable to produce the dead fish. Eels in the Esopus moved in quickly following a fish kill, gorging themselves on the refuse, in effect destroying valuable evidence. J. Douglas Sheppard, discussion with the author, July 8, 2008.

86. "DEC Moves against NYC for Fish Killed in Ulster," *Times Herald Record*, November 2, 1974.

87. "Summary of New York City Reservoir Releases Problem in Upper Delaware Basin," internal document of Department of Environmental Conservation, November 19, 1975, series L0131–78, box 1, folder 8, NYSA.

88. Chuck Schwartz, president, Phoenicia Fish and Game Association, to Cyril Moore, an attorney employed by New York State, December 8, 1974, series 17067–02, New York State Department of Environmental Conservation, Office of Hearings and Mediation Services, Decisions, Orders, Rulings, and Hearing Reports, 1987–2002, box 3, folder 8 (Esopus Creek Fishkill), NYSA.

89. "DEC Moves against NYC," *Times Herald Record*.

90. The overlap between natural and human processes is central to environmental history. For a good example of this in a riparian context, see Kelman, *River and Its City*, 50–69.

91. White, *Organic Machine*.

92. Scott, *Seeing Like a State*, 262–306. Scott's book is replete with examples of the unfortunate consequences of the simplification of nature by governments of various ideological persuasions.

93. Department of Environmental Conservation, press release, June 22, 1976, series 16841–92, box 13, folder 14, NYSA.

94. Catskill Waters, Public Statement Concerning Downstream Water Releases, May 7, 1976, series 16841–92, box 13, folder 15, NYSA.

95. Untitled document by New York State Department of Environmental Conservation, 1976, series 16841–92, box 10, folder 13, NYSA.

96. John McPhee identified the raising of expectations among Mississippi River residents sparked by increasingly interventionist actions by the Army Corps of Engineers. "But now that Old River is valved and metered, there are two million nine hundred thousand potential complainers, very few of whom are reluctant to present a grievance to the Corps." McPhee, *Control of Nature*, 22.

97. Stutz, *Natural Lives, Modern Times*, 342.

98. For an overview of the development of New York State's approach to environmental issues, see Birkland et al., "Environmental Policy in New York." Despite what appeared to be a newfound state resolve, many DEC employees viewed their superiors as timid in the face of resistance by the city. State biologists resented the so-called stipulation agreement, a settlement between the city and state reached in 1980, which they viewed as undermining many of the gains they had made in recent years. Sheppard, discussion.

99. Stock and Johnston, introduction in Stock and Johnston, *Countryside in the Age of the Modern State*, 5–6.

100. Statement by Einar Eklund at public hearing held in Oneonta, NY, March 9, 1977, series L0131–78, Hearing Files and Transcripts, 1975–77, Legislative Assembly, Standing Committee on Environmental Conservation, box 1, folder 5, NYSA.

101. Mark Milani to William Clarke, deputy chief permit administrator, February 26, 1987, series 17067–02, box 1, folder "Legisl. Hearing," NYSA.

102. Catskill Waters brought together fishermen from the entire Delaware River basin. While Catskill residents served as the public face of the organization and did much of the lobbying, fishermen from New York City who were members of the Theodore Gordon Flyfishers underwrote much of the effort. Sheppard, discussion.

103. Van Put, *Trout Fishing in the Catskills*, 389–91.

104. Ibid., 390–92.

105. Department of Environmental Conservation, press release, June 22, 1976.

106. John Hoeko, president Catskill Waters, "Public Statement concerning Downstream Water Releases," May 7, 1976, series 16841–92, box 13, folder 15, NYSA.

107. Frank Mele, vice president, Catskill Waters, "Public Statement concerning Downstream Water Releases," May 7, 1976, series 16841–92, box 13, folder 15, NYSA.

108. Smiley, *Catskill Crafts*, 21.

109. Van Put, *Trout Fishing in the Catskills*, 392.

110. Ibid.

111. Smiley, *Catskill Crafts*, 22.

112. Birkland et al., "Environmental Policy in New York," 396.

113. Francis, *Land of Little Rivers*, 97.

114. Michael Longuil to Peter Berle, commissioner, New York State Department of Environmental Conservation, September 12, 1977, series 16841–92, box 13, folder 8, NYSA.

115. Reservoir Release Task Force to Governor Hugh Carey, December 22, 1977, series 16841–92, box 13, folder 8, NYSA.

116. J. Douglas Sheppard, *New York Reservoir Releases Monitoring and Evaluation Program—Delaware River—, Technical Report No. 83–5*, 38.

117. Ibid., 32.

118. Capossela, *Good Fishing in the Catskills*, 173.

119. Francis McArdle to Mayor Ed Koch, May 15, 1978, Mayor Ed Koch Departmental Correspondence, box 141, folder 1, 1978 (Environmental Protection), MACNY.

120. Flippen, *Nixon and the Environment*, 227–28. On the collision of different eras of natural resource policy, see Klyza and Sousa, *American Environmental Policy*, chap. 2.

121. For a brief history of the DEC, see www.dec.ny.gov/about/9677.html; see also Birkland et al., "Environmental Policy in New York," 396–400.

122. Sociologist Michael Mann distinguishes between two types of state power, despotic power and infrastructural power. Despotic power denotes the capacity of state elites to "rule unchecked by other centers of power or by civil society." Infrastructural power refers to the state's ability to implement policies throughout a given region. New York City deployed both forms of power in constructing its water system. For a discussion of how Mann's typology sheds light on the American state, see Novak, "Myth of the 'Weak' American State," 763–64.

123. Serio, discussion.

124. Andrews, *Managing the Environment*, 233.

125. When he signed the National Environmental Policy Act into law on live television, President Nixon dubbed the 1970s the "decade of the environment."

6. The Rise of Watershed Management

1. Allan Gold, "2 Boroughs Face a Ban on Building," *New York Times*, June 11, 1990; Appleton describes these challenges in "Post-Delaware Era," 216–35.

2. Robert F. Kennedy Jr., interview by Nancy Burnett, *BTS*, July 12, 2002, 78.

3. Appleton, "Regional Landscape," 13.

4. National Research Council, *Watershed Management for Potable Water Supply*, 107–15.

5. Tarr, *Search for the Ultimate Sink*, 192–95.

6. The city of Seattle owns virtually all the land in its watershed; the federal government owns the vast majority of the watersheds that supply San Francisco and Portland. For brief snapshots of these other unfiltered water systems, see Finnegan, "New York City's Watershed Agreement," 579.

7. For background on the protracted battle to build the Croton plant, see Environmental Protection Agency, "EPA Order against NYC Points Out Need to Filter Croton Supply"; Freud, *Why New York City Needs a Filtered Croton Supply.*

8. As Commissioner Appleton observed, "The City relied on the size of its watersheds and the volume of water in its reservoirs to dilute any pollution down to the point of being harmless." Appleton, "Regional Landscape Preservation," 9.

9. Robert F. Kennedy Jr. interview, *BTS,* 79.

10. Appleton, "Regional Landscape Preservation," 13.

11. The desire to protect New York City's Catskill water supply led to the passage of a state law in 1915 that granted the city broad rights to proscribe and limit a wide range of activities. This original set of regulations barred privies and stables near watercourses or reservoirs and prohibited swimming in reservoirs. See Galusha, *Liquid Assets,* 151. On the draft regulations and the initial response in the Catskills, see Allan R. Gold, "New York's Water Rules Worry Catskills," *New York Times,* September 17, 1990.

12. Alan Rosa, interview by Nancy Burnett and Virginia Scheer, *BTS,* February 6, 1996, 133.

13. Albert Appleton, discussion with the author, New York, NY, April 9, 2008.

14. Kenneth Markert, interview by Nancy Burnett and Virginia Scheer, *BTS,* January 18, 1996, 86.

15. Jeffrey S. Baker, interview by Nancy Burnett, *BTS,* Albany, NY, December 15, 1998, 4–5.

16. Daniel Ruzow, interview by Nancy Burnett, *BTS,* August 2, 2001, 107–8.

17. Anthony Bucca, interview by Nancy Burnett and Virginia Scheer, *BTS,* February 15, 1996, 37.

18. Markert, interview, *BTS,* 85.

19. Ibid., 97.

20. Dennis Rapp, discussion with the author, February 20, 2008. The regulations were drafted primarily by two DEP lawyers.

21. Baker, interview, *BTS,* 4.

22. Appleton, "Regional Landscape Preservation," 15.

23. Appleton, discussion.

24. Markert, interview, *BTS,* 99.

25. Appleton, "Regional Landscape Preservation," 10–15; Appleton, discussion; David Warne, discussion with the author, January 13, 2009. Increasing interest in urban conservation was an important trend both nationally and internationally. See Platt, Rowntree, and Muick, *Ecological City.*

26. Rapp, discussion.

27. Rapp, discussion.

28. Ad Hoc Task Force on Agriculture and New York City Watershed Regulations, *Ad Hoc Task Force on Agriculture and New York City Watershed Regulations: Policy Group Recommendations,* 3.

29. Walter and Walter, "New York City Watershed Agricultural Program," 6.

30. Martin, *Water for New York,* 135.

31. Scott, *Seeing Like a State,* 11–52. Scott explores the simplifying effects of mapping projects in early nation states. In the case of the WFP, detailed maps of individual farms represented an effort to get beyond abstraction and use maps as a tool for ecologically sensitive farm management.

32. Rapp, discussion.

33. Rapp, discussion; Appleton, "Post-Delaware Era," 228–30.

34. Walter and Walter, "New York City Watershed Agricultural Program," 6.

35. Tara Collins and Fred Huneke, discussion with the author, May 24, 2011.

36. Albert Appleton, *How New York City Used an Ecosystem Services Strategy*, 8.

37. Appleton, "Regional Landscape Preservation," 10.

38. Watershed Agricultural Council, "Davis Farm," http://nycwatershed.org/clw_farm_archive_davis.html (accessed July 2007).

39. Proof of these improvements was the removal of the Cannonsville Reservoir from a state list of reservoirs that habitually exceeded permitted phosphorous levels. Thomas Snow, state coordinator New York State Watershed, discussion with the author, August 25, 2008. Mark Sagoff highlights the intuitive appeal of the Catskill watershed protection program for many environmentalists eager to demonstrate the economic value of the services provided by nature. See Sagoff, "On the Value of Natural Ecosystems."

40. Appleton, "Post-Delaware Era," 229. On the need for new strategies to combat nonpoint pollution, see Sabatier, Weible, and Ficker, "Eras of Water Management in the United States," 41–42; Lubell et al., "Watershed Partnerships and the Emergence of Collective Action Institutions."

41. Andrews, *Managing the Environment*, 352.

42. The involvement of many stakeholders and the recognition of the value of local knowledge are two of the defining features of this collaborative approach. See Sabatier, Weible, and Ficker, "Eras of Water Management in the United States," 47–51; Brunner et al., *Adaptive Governance*, vi–xi; John, *Civic Environmentalism*; Wondolleck and Yaffee, *Making Collaboration Work*. Clinton appointees Michael Dombeck, chief of the Forest Service, and Bruce Babbitt, secretary of the Interior, both encouraged the use of collaborative approaches. See Dombeck, Wood, and Williams, "Focus: Restoring Watersheds," 26.

43. Rapp, discussion.

44. Appleton, "Post-Delaware Era," 228–29.

45. Appleton, discussion.

46. Michael Specter, "Reprieve for New York on Water Filtering," *New York Times*, January 20, 1993.

47. K. O. Wilson, "'The Avenue of Attack That Will Kill Us': Condemnation Is New York City's Intent," *Delaware County Times*, February 8, 1994; K.O. Wilson, "Robert Kennedy Says Condemnation Now!" *Delaware County Times*, February 22, 1994.

48. Keith Porter, interview by Nancy Burnett, *BTS*, July 20, 2001, 93.

49. Robert F. Kennedy Jr. called for extensive land acquisition in the Catskills to safeguard water quality. See Wilson, "Robert Kennedy Says Condemnation Now!"; K. O. Wilson, "Irreconcilable Differences," *Delaware County Times*, March 22, 1994.

50. For a comprehensive discussion of the new conservation measures, see Hazen and Sawyer, "The New York City Water Conservation Program," series 10767–02, box 3, NYSA.

51. Mayor's Intergovernmental Task Force on New York City Water Supply Needs, *Increasing Supply, Controlling Demand: Interim Report*.

52. DEC Assistant Counsel Carl Dworkin to Administrative Law Judge Susan DuBois ("Reply to Briefs of Conservation Issues"), April 13, 1990, series 10767–02, box 2, folder Corr. 2/19/02–6/18/02, NYSA.

53. News clippings, series 10767–02, box 2, folder Corr. 2/19/02–6/18/02, NYSA.

54. Judd, *How Much Is Enough?* 35.

55. Appleton, discussion.

56. A 1989 study identified toilets as the primary cause of indoor water waste. Hazen and Sawyer, "New York City Water Conservation Program," 3–6.

57. Warren Liebold, discussion with the author, January 18, 2007; Mitchell, *Catskills*, 103.

58. Liebold, discussion.

59. New York City Department of Environmental Protection, *A Handy Resource Guide to the New York City Toilet Rebate Program*, 2nd ed., 1–3; Liebold, "Conservation Program Development in New York City," 3.

60. Bureau of Water and Energy Conservation, *Water-Saving Plumbing Fixture Retrofit and Rebate Program*, 8.

61. New York City Department of Environmental Protection, *Water Conservation Program*, 16–17.

62. Liebold, discussion.

63. Judd, *How Much Is Enough?* 13.

64. The analogy between the city's success in reducing crime and its water conservation efforts comes from Sam Roberts, "More Masses Are Huddling in New York, but They're Using Less Water," *New York Times*, October 3, 2006.

65. The literature on both crime and gentrification in New York City is vast. Two useful studies are Blumstein and Wallman, *Crime Drop in America*; and Siegel, *Prince of the City*.

66. Bureau of Water and Energy Conservation, *Water-Saving Plumbing Fixture Retrofit and Rebate Program*, 8; Albert Appleton, "Financial Tools, Financial Strategy and Water Utility Operating and Environmental Performance," 11. Ironically, a budget crunch led to the cancellation of most of the festivities the city had planned to commemorate the 150th anniversary of the Croton system. See Koeppel, *Water for Gotham*, xii.

67. Liebold, discussion.

68. Marilyn Gelber, interview by Nancy Burnett, *BTS*, July 6, 2001, 53.

69. Ibid., 53–54.

70. Shelton, interview, *BTS*, 157–58.

71. K. O. Wilson, "Sewage in New York City Drinking Fountains?" *Delaware County Times*, September 23, 1994.

72. Ruzow, interview, *BTS*, 114–115.

73. Joyce Purnick, "Not Party, but Need Led Mayor," *New York Times*, October 27, 1994.

74. Bucca, interview, *BTS*, 30.

75. Cronin and Kennedy, *Riverkeepers*, 224.

76. Erin Crotty, interview with Nancy Burnett, *BTS*, August 22, 2001, 33.

77. Bucca, interview, *BTS*, 27.

78. Ruzow, interview, *BTS*, 116.

79. Guy Olivier Faure and Jeffrey Rubin, eds., *Culture and Negotiation: The Resolution of Water Disputes*.

80. Crotty, interview, *BTS*, 36.

81. Finnegan, "New York City's Watershed Agreement," 624.

82. Ibid.

83. Gelber, interview, *BTS*, 57; Ruzow, interview, *BTS*, 122.

84. Bucca, interview, *BTS*, 43.

85. Gelber, interview, *BTS*, 57.

86. Bucca, interview, *BTS*, 25.

87. Gelber, interview, *BTS*, 61.

88. On the demise of scientific management, see Brunner et al., *Adaptive Governance*, chap. 1. On the rise of a more consensual mode of environmental governance, see Klyza and Sousa, *American Environmental Policy*, 195–246.

89. Rosa, interview, *BTS*, 141.

90. Cronin and Kennedy, *Riverkeepers*, 225.

91. Gelber, interview, *BTS*, 66.

92. At the city's insistence, the Croton watershed was also included in the MOA. New York began to plan construction of a filtration plant for Croton water, but it adopted a two-pronged strategy of pollution prevention and pollution cleanup for its Croton supply.

93. Catskill Watershed Corporation, *Summary Guide to the Terms of the Watershed Agreement*; New York City Department of Environmental Protection, *New York City Watershed Memorandum of Agreement*.

94. Catskill Watershed Corporation, *Summary Guide to the Terms of the Watershed Agreement*, 10–11.

95. Finnegan, "New York City's Watershed Agreement," 625.

96. Catskill Watershed Corporation, *Summary Guide to the Terms of the Watershed Agreement*, 13–18.

97. Ibid., 19–23.

98. Ibid., 23.

99. Gelber, interview, *BTS*, 59.

100. Ruzow, interview, *BTS*, 126–27.

101. "At Last, a Watershed Agreement," *New York Times*, November 3, 1995.

102. Markert, interview, *BTS*, 98.

103. Warren Liebold emphasized the sincere embrace of conservation by DEP employees, but acknowledged that the threat of state action accelerated conservation efforts and enabled the DEP rank and file to secure the backing of agency leaders for a vigorous conservation program. Liebold, discussion.

104. Secretary of the Interior Babbitt played a leading role in institutionalizing more consensual approaches to environmental protection. See Babbitt, *Cities in the Wilderness*. On gridlock, the reliance on courts, and habitat conservation plans, see Klyza and Sousa, *American Environmental Policy*.

105. In the early 2000s, historians proposed a revised conception of American politics that blends institutional insights—premised on the idea that the structure and mechanics of government power shape political outcomes—with an emphasis on contingency and resistance by private interests. Known as the new political history, this approach seeks to reconcile the findings of institutional scholars with strong evidence of countervailing pressure from the private sector. Studies in a new political history vein include Jacobs, Novak, and Zelizer, *Democratic Experiment*; and Klein, *For All These Rights*.

106. Liebold, discussion.

7. Implementing the Watershed Agreement

1. The EPA previously granted the city short-term waivers, formally known as FADs, Filtration Avoidance Determinations. The 1997 waiver was qualitatively different because it endorsed the programs and institutional mechanisms contained in the MOA as the basis of a long-term watershed protection model.

2. In most basins, the city acquired approximately 20 percent of the properties it sought to purchase. For data by basin, see New York City Department of Environmental Protection, Bureau of Water Supply, *Long-Term Land Acquisition Plan, 2012 to 2022*, 26–29.

3. Steve Justice, "Watershed Signs Irk Delaware Leaders," *Daily Star*, March 18, 1999.

4. Ibid.

5. Alan Rosa, "Executive Director's Report," *Watershed Advocate*, Spring 1999, 2. (The *Advocate* was the quarterly newsletter of the Catskill Watershed Corporation.)

6. Steve Justice, "Delaware Asks NYC to Help with Stream Work," *Daily Star*, March 25, 1999; Steve Justice, "Watershed MOA a Success So Far, Most Officials Say," *Daily Star*, April 3, 1999.

7. "Watershed MOA a Success So Far"; on Miele, see Kennedy, Sullivan, and Postman, *Watershed for Sale*, 4.

8. "Friendlier Signs," *Watershed Advocate*, Fall 2000, 7.

9. "All for Water," *Watershed Advocate*, Summer 2002, 5; "Gone, but Not Forgotten," *Watershed Advocate*, Summer 2004, 8.

10. Berman, *All That Is Solid Melts into Air*.

11. "$310,000 in Economic Development Grants Awarded," CWC press release, June 1, 2004, www.cwconline.org/news/press/2004/2004_0601.htm (accessed May 2011).

12. T. M. Bradshaw, "New Life for Old Visions," *Catskill Mountain Region Guide*, January 2007, www.catskillmtn.org/guide-magazine/articles/2007–01-new-life-for-old-visions.html (accessed June 2009).

13. "From Frustration to Joy: Amy Jackson's Triumph," *Watershed Advocate*, Summer 2006, 1–2.

14. Collins and Huneke, discussion.

15. Jeff Baker (attorney, Coalition of Watershed Towns), discussion with the author, May 26, 2011.

16. Alan Rosa, discussion with the author, May 25, 2011.

17. New York City Department of Environmental Protection, Bureau of Water Supply, *Long-Term Land Acquisition Plan*, 8.

18. Rosa, discussion.

19. Holly Aplin, letter to the editor, "NYC Restrictions Are Hurting Area," *Daily Star*, March 26, 2002.

20. Rosa, discussion.

21. "New York City Opens Trails in Watershed," *Daily Star*, June 4, 1999.

22. "New York City to Open Watershed Lands for Deer Hunting and Establish Bow-Hunting Season; Permits Required," www.nyc.gov/html/dep/html/press_releases/10–99pr.shtml (accessed December 2010).

23. Rosa, discussion.

24. Jack McShane and Alan White, discussion with the author, February 22, 2006.

25. Snow, discussion.

26. Patricia Breakey, "DEP Expands Land-Use Policies," *Daily Star*, August 15, 2006.

27. New York City Department of Environmental Protection, "Proposed Amended Rules to Expand Recreational Use in Watershed," http://home2.nyc.gov/html/dep/html/press_releases/08–16pr.shtml, accessed June 2011; Rosa, discussion.

28. Rosa, discussion.

29. Anne Schwartz, "The Gates and the Water Filtration Plant," *Gotham Gazette*, February 18, 2005, www.gothamgazette.com/article/parks/20050218/14/1328 (accessed September 2008).

30. Joyce Purnick, "Water Plant with a Spoon of Honey," *New York Times*, May 8, 2003; New York City Department of Environmental Protection, *Public Hearings—Croton Water Treatment Plant: Environmental Impact Statements*, vol. 5 ; "Water Plant Making Parks Greener," *Metro New York City Parks*, September 2007, 2. Some park advocates questioned whether the Parks Department used Croton funds to substitute for other capital funds intended for Bronx parks. They also criticized the protracted timeline for spending Croton funds. See Daniel Beekman, "City Controller John Liu Launches Audit of Capital Spending on Parks, Including Controversial Croton Funds for Bronx," *New York Daily News*, February 7, 2012.

31. Purnick, "Water Plant."

32. "A Bronx Story: Four Bronx Parks Receive $14 Million in Renovations," New York City, www.nycgovparks.org/sub_newsroom/press_releases/press_releases.php?id=20532 (accessed August 2008).

33. "DEP Expands Access for Kensico and New Croton Reservoirs," www.nyc.gov/html/dep/html/press_releases/10–102pr.shtml (accessed June 2011).

34. Nate Schweber, "In the Suburbs, When the Cold Is Biting, So Are the Fish," *New York Times*, February 14, 2011.

35. For the original vision of individualized plans for mitigating agricultural solution, see Ad Hoc Task Force on Agriculture; a comprehensive description of Whole Farm Planning is available at the Watershed Agriculture Council's website, www.nycwatershed.org/ag_planning.html (accessed June 2011).

36. "Wastewater Treatment Plant Upgrades Moving Ahead," www.nyc.gov/html/dep/html/press_releases/01–40pr.shtml (accessed June 2001); U.S. Environmental Protection Agency, *Assessing New York City's Watershed Protection Program*, 132–44.

37. U.S. Environmental Protection Agency, *Assessing New York City's Watershed Protection Program*, 142–45.

38. U.S. Environmental Protection Agency, *New York City Filtration Avoidance Determination*, 5.

39. New York City Department of Environmental Protection, *2006 Long-Term Watershed Protection Program*, 26.

40. Catskill Watershed Corporation, *2009 Annual Report*, 2.

41. Catskill Watershed Corporation, *Board of Directors Meeting Minutes*, February 2, 2010, www.cwconline.org/about/ab_board.html (accessed May 2011).

42. Drew Harty and Nelson Bradshaw, directors, *Of Streams and Dreams: The Programs of the Catskill Watershed Corporation*.

43. New York City Department of Environmental Protection, *2006 Long-Term Watershed Protection Program*, 11.

44. McShane and White, discussion.

45. Rosa, discussion. Intensive analysis of water quality throughout the watersheds is a critical component of source protection efforts. See Stroud Water Research Center, *Water Quality Monitoring in the Source Water Areas for New York City*.

46. Baker, discussion.

47. New York City Department of Environmental Protection, *Final Environmental Impact Statement (EIS): The Extended New York City Land Acquisition Program*, ES-2.

48. New York City Department of Environmental Protection, Bureau of Water Supply, *Long-Term Land Acquisition Plan*, 3–4.

49. For case studies that examine the balance between efficiency and equity in water resource allocation, see Whiteley, Ingram, and Perry, *Water, Place, and Equity*.

50. New York City Department of Environmental Protection, Bureau of Water Supply, *Long-Term Land Acquisition Plan*, 7–8.

51. McShane and White, discussion.

52. New York City Department of Environmental Protection, Bureau of Water Supply, *Long-Term Land Acquisition Plan*, 7, 12.

53. Kennedy, Sullivan, and Postman, *Watershed for Sale*, 2.

54. New York City Department of Environmental Protection, Bureau of Water Supply, *Long-Term Land Acquisition Plan*, 7, 12; New York City Department of Environmental Protection, *Final Environmental Impact Statement*, ES-4.

55. "Proposed Amended Rules to Expand Recreational Use in Watershed," www.nyc.gov/html/dep/html/press_releases/08–16pr.shtml (accessed June 2011).

56. New York City Department of Environmental Protection, Bureau of Water Supply, *Long-Term Land Acquisition Plan*, 8; Dean Frazier and Tom Hilson, discussion with the author, May 26, 2011.

57. Mark Boshnack, "NYC Gets Plan OK," *Daily Star*, July 31, 2007.

58. Downeast Development Consulting Group, *New York City Watershed Economic Impact Assessment Report*, 135–43; Frazier and Hilson, discussion.

59. Downeast Development Consulting Group, *New York City Watershed Economic Impact Assessment*, 143.

60. New York City Department of Environmental Protection, Bureau of Water Supply, *Long-Term Land Acquisition Plan*, 8.

61. Frazier and Hilson, discussion.

62. Ibid.

63. Editorial, "A Plan to Protect New York Water," *New York Times*, February 25, 2007.

64. "EPA Grants NYC New Waiver from Filtering Drinking Water from Its Catskill/Delaware System," http://yosemite.epa.gov/opa/admpress.nsf/3881d73f4d4aaa0b852573590 03f5348/54aeb32b2719f5f585257328004c70da!OpenDocument (accessed July 2011); Patricia Breakey, "Delaware Officials Object to Proposal for 10-Year Waiver for Filtering Water," *Daily Star*, April 14, 2007; Anthony DePalma, "City's Catskill Water Gets 10-Year Approval," *New York Times*, April 13, 2007.

65. DePalma, "City's Catskill Water Gets 10-Year Approval"; Carl Campanile, "NYC Ducks $8B Soaking," *New York Post*, April 13, 2007.

66. Frazier and Hilson, discussion.

67. "NY Watershed Towns' Lawsuit Fails," *AWWA Streamlines* 20 (January 20, 2009), www.awwa.org/publications/StreamlinesArticle.cfm?itemnumber=44937 (accessed June 2011).

68. Baker, discussion.

69. For background on the Environmental Quality Review Act, see Department of Environmental Conservation website, www.dec.ny.gov/permits/357.html (accessed July 2011); Baker, discussion.

70. Julia Reischel, "New Tax Scheme at Center of Watershed Deal," *Watershed Post* (blog), November 15, 2010, www.watershedpost.com/2010/new-tax-scheme-center-watershed-deal (accessed November 2011).

71. New York City Department of Environmental Protection, "State, City Announce Landmark Agreement to Safeguard New York City Water," www.nyc.gov/html/dep/html/press_releases/11–11pr.shtml (accessed March 2011); Reischel, "New Tax Scheme."

72. Baker, discussion.

73. Baker, discussion; New York City Department of Environmental Protection, "State, City Announce Landmark Agreement."

74. New York City Department of Environmental Protection, "DEP Expands Access for Kensico and New Croton Reservoirs," www.nyc.gov/html/dep/html/press_releases/10–102pr.shtml (accessed June 2011); New York City Department of Environmental Protection, "DEP Plans to Open 6,600 More Watershed Acres for Recreation," www.nyc.gov/html/dep/html/press_releases/11–34pr.shtml (accessed May 2011).

75. New York City Department of Environmental Protection, *2006 Long-Term Watershed Protection Program*, 23.

76. Rosa, discussion.

77. New York City Department of Environmental Protection, "DEP Unveils Design to Repair Leaks in the 85-Mile Delaware Aqueduct," www.nyc.gov/html/dep/html/press_releases/10–99pr.shtml (accessed December 2010).

78. New York State Comptroller, *Delaware Aqueduct System*, 2–3; Kennedy et al., *Finger in the Dike*, 17–39.

79. Riverkeeper, "Memorandum of Support for S6276 and A10140," on Riverkeeper website, www.riverkeeper.org/ . . . /RvK-Memorandum-of-Support-A10140-Wawarsing-Flooding-1-pdf (accessed June 2011).

80. Adam Bosch, "NYC Chips in $3.7 M to Buy Wawarsing Homes," *Times Herald-Record,* June 25, 2011.

81. New York State Comptroller, *Delaware Aqueduct System*, 8.

82. New York City Department of Environmental Protection, "DEP Unveils Design to Repair Leaks."

83. For a good summary of the project, see "New York City Tunnel No. 3 Construction, USA," on Water-Technology.net website, www.water-technology.net/projects/new-york-tunnel-3/ (accessed July 2011).

84. Arthur, Bohm, and Layne, "Hydraulic Fracturing Considerations for Natural Gas Wells of the Marcellus Shale," www.dec.ny.gov/docs/materials_minerals_pdf/GWPC-Marcellus.pdf (accessed June 2011).

85. America's Natural Gas Alliance, "Hydraulic Fracturing: How It Works," www.anga.us/learn-the-facts/hydraulic-fracturing-101 (accessed July 2011).

86. Howarth, Santoro, and Ingraffea, "Methane and the Greenhouse-Gas Footprint of Natural Gas from Shale Formations."

87. A typically skeptical article is Bateman, "Colossal Fracking Mess"; Fox, *Gasland*.

88. Hazen and Sawyer, *Final Impact Assessment Report*.

89. Frazier and Hilson, discussion.

90. Mireya Navarro, "State Decision Blocks Drilling for Gas in Catskills," *New York Times*, April 23, 2010; Sinding and Goldstein, "Natural Gas Drilling Threatens NYC Water Supply," www.crainsnewyork.com/article/20110717/SUB/307179981 (accessed July 2011).

91. Platt, "Water Supply Protection through Watershed Management." The Massachusetts Department of Conservation and Recreation manages the state's watershed programs and works closely with watershed property owners, but there is no Massachusetts equivalent to the MOA.

92. Rosa, discussion.

93. Rosa, discussion; Frazier and Hilson, discussion.

94. Rosa, discussion.

95. Layzer, *Natural Experiments*.

96. On the environmental benefits of urban living, see Owen, *Green Metropolis*.

Epilogue

1. Rosa, discussion.

2. Collins and Huneke, discussion.

3. Environmental Law Institute, *NEPA Success Stories*.

4. Collins and Huneke, discussion.

5. On REDD, see "REDD May Yet Survive Copenhagen Failures," www.carbonpositive.net/viewarticle.aspx?articleID=1786 (accessed July 2011).

6. Collins and Huneke, discussion.

7. Watershed Agricultural Council, "International Visitors Learn from Watershed Farmsteads," *Watershed Farm and Forest*, Winter 2009, 1–2.

8. Collins and Huneke, discussion.

9. Rymer, "Reuniting a River"; Layzer, *Natural Experiments*.

10. Collins and Huneke, discussion.

11. Melnick, *Taking Stock*, 161.

12. Klyza and Sousa, *American Environmental Policy*, chap. 2; Layzer, *Natural Experiments*; Babbitt, "Diversity."

13. New York City Department of Environmental Protection, "NYC to Acquire 1,655 Acres of Land for Watershed Protection," www.nyc.gov/html/dep/html/press_releases/11–75pr.shtml (accessed September 2011), and "DEP Completes Ashland Wastewater Treatment Plant," www.nyc.gov/html/dep/html/press_releases/11–76pr.shtml (accessed September 2011).

14. Warne, discussion.

Bibliography

Primary Sources

Manuscript Collections

George Bliss Agnew Papers, Manuscripts and Archives Division, New York Public Library, New York, NY

John Shaw Billings Records, Manuscripts and Archives Division, New York Public Library, New York, NY

Herbett Howlett Papers, Special Collections and University Archives, Rutgers University Libraries, New Brunswick, NJ

Chet Lyons Collection, Olive Free Library Archives, Olive, NY

Robert Moses Papers, Manuscripts and Archives Division, New York Public Library, New York, NY

New York Board of Water Supply Transcripts Relating to the Pepacton and Cannonsville Reservoirs, Paul Cooper Jr. Archives, Stevens-German Library, Hartwick College, Oneonta, NY

New York City Parks Department Library, Vertical Files, New York, NY

Olive Free Library Archives, Olive, NY

Nelson A. Rockefeller Gubernatorial Papers, Rockefeller Archive Center, Sleepy Hollow, NY

Abel Wolman Papers, Special Collections, Milton S. Eisenhower Library, Johns Hopkins University, Baltimore, MD

Town of Olive Archives, Olive, NY

John Watson Photograph Collection
Robert Pleasants Collection
Vera Sickler Papers

Massachusetts Institute of Technology, Institute Archives, Cambridge, MA

Desmond Fitzgerald Papers
John Ripley Freeman Papers

New York State Archives, Albany, NY

Division Director's Subject and Correspondence Files, 1962–96, Division of Fish, Wildlife, and Marine Resources
Drainage Basin Subject Files, New York State Department of Environmental Conservation
Executive Office, Commissioners' and Deputy Commissioners' Correspondence, Subject Files, and Orders, 1948–95, New York State Department of Environmental Conservation
Hearing Files and Transcripts, 1975–77, New York State Legislature (Assembly), Standing Committee on Environmental Conservation
Interstate Basin Commission Administration Files
Office of Hearings and Mediation Services, Decisions, Orders, Rulings, and Hearing Reports, 1987–2002, New York State Department of Environmental Conservation, Town of Olive Records

Collections on Microfilm

Ashokan Reservoir Awards

Municipal Archives of the City of New York

Mayor Vincent Impelliterri Subject Files
Mayor Ed Koch Departmental Correspondence
Mayor George McClellan Papers
Mayor William O'Dwyer Subject Files
Mayor Robert Wagner Subject Files
Department of Parks and Recreation Records

Loeb Library, Harvard University

Vertical Files

Westchester County Archives, Elmsford, New York

Carl Pforzheimer Papers
Environmental Health Services Annual Reports

Newspapers and Magazines

American City Magazine
Brooklyn Daily Eagle
Brooklyn Daily Times
Catskill Center News

Century Illustrated Magazine
The Clarion, New York, NY
The Daily Plant, New York, NY
Daily Star, Oneonta, New York
Delaware County Times
Delaware Republican Express
Gotham Gazette
Harper's Magazine
Harper's Weekly
The Independent . . . Devoted to the Consideration of Politics, Social and Economic Tendencies, History, Literature, and the Arts
Kingston Daily Freeman, Kingston, NY
Life Magazine
Magazine of the New York Philanthropic League
Metropolis
Mount Kisco Recorder, Mount Kisco, NY
Munsey's Magazine
New Rochelle Pioneer, New Rochelle, NY
Newsday
Newsweek
New York Daily News
New York Evening Post
New York Herald
New York Herald Tribune
New York Observer
New York State Conservationist
New York Sun
New York Times
New York Tribune
New York World
New York World-Telegram and the Sun
New Yorker
The Outlook
Philadelphia Evening Bulletin
Philadelphia Inquirer
Pine Hill Sentinel, Pine Hill, NY
Scientific American
Scientific American Supplement
Times Herald Record, Middletown, NY
Ulster: A Regional Magazine
Westchester News
Woodstock Times, Woodstock, NY

Published Government Documents

Ad Hoc Task Force on Agriculture and New York City Watershed Regulations. *Ad Hoc Task Force on Agriculture and New York City Watershed Regulations: Policy Group Recommendations.* New York: New York City Department of Environmental Protection, 1991.

Board of Aldermen and Common Council of the City of New York, Committee on County Affairs. *In the Matter of Proposed Removal of 42nd Street Reservoir, Memorandum in Opposition Thereto, Filed by New York Board of Fire Underwriters.* New York: Douglas Taylor & Co., Printers, 1896.

Board of Aldermen. *Proceedings of the Board of Aldermen of the City of New York,* October 3–December 20, 1905. New York: Board of Aldermen, 1905.

———. *Proceedings of the Board of Aldermen of the City of New York,* January 1–March 19, 1912. Vol. 1. New York: Martin B. Brown Press, 1912.

Board of Water Supply of the City of New York. *Annual Reports of the Board of Water Supply of the City of New York.* New York: Board of Water Supply, 1907–73.

———. *Catskill Water Supply: A General Description.* New York: Board of Water Supply, 1915–19.

———. *Catskill Water System News.* New York: Board of Water Supply. June 20, 1911–December 20, 1913.

———. *City of New York Additional Water Supply, Catskill Aqueduct, Inauguration of Construction, Peekskill, NY, June 20, 1907.* New York: Board of Water Supply, 1907.

———. *Delaware Water Supply News.* New York: Board of Water Supply, August 1938–October 1965.

———. *Minutes of the Meetings.* New York: Herald Square Press, 1920.

———. *Minutes of the Meetings.* New York: Herald Square Press, 1922.

———. *1,820,000,000 Gallons per Day: 50th Anniversary of the Board of Water Supply.* New York: Board of Water Supply, 1955.

———. *Opinion of Hon. John W. Davis Relating to Right of the City of New York to Utilize Certain Waters of the Delaware River, June 22, 1928.* New York: Board of Water Supply, 1928.

———. *Report of the Board of Water Supply of the City of New York to the Board of Estimate and Apportionment, July 27, 1927.* 1927. Reprint, New York: Board of Water Supply, 1937.

———. *Report of the Board of Water Supply to the Mayor of the City of New York on Completion of the First Stage of the Catskill Water Supply* System. New York: Board of Water Supply, 1917.

Buerle, David E. *Water Resource Issues of the Catskill Region: Identification, Analysis, and Recommendations.* Stamford, NY: Temporary State Commission to Study the Catskills, 1975.

Burch, Charles Newell. *State of New Jersey, Plaintiff, vs. State of New York and City of New York, Defendants.* Washington, DC: 1931.

Bureau of the Budget. *Study of Water Conservation—Metering as a Basis for Charges for the Use of Water.* New York: Bureau of the Budget, 1948.

Carman, Simon. *Water Supply Problems, Westchester County, Preliminary Report 1955.* White Plains, NY: Westchester County, 1955.

Chapin, Alfred C. *Address to the Common Council of Brooklyn, July 11, 1889: The Water Supply, Its Proposed Increase, the Money Available for Improved Pavements, New Sewers, Relief Sewers, Additional Street Cleaning, New Buildings and More School Houses.* Brooklyn, 1889.

Commissioner of Public Works. *History and Description of the Water Supply of the City of Brooklyn.* Brooklyn: Commissioner of Public Works, 1896.

Committee on Additional Water Supply. *Report of Mayor O'Brien's Committee on Additional Water Supply.* New York, 1933.

Cooper, Linda. *Walker's Guide to the Old Croton Aqueduct.* Staatsburg, NY: New York State Office of Parks, Recreation and Historic Preservation, Taconic Region, in cooperation with the Lucius N. Littauer Foundation, 1986.

Corporation Counsel of the City of New York. *In the Matter of the Application of the City of New York to the Water Power and Control Commission for the Approval of Its Maps, Profile and Plan for Securing an Additional Supply of Water from the West Branch of the Delaware River: Memorandum of the City of New York.* New York: Sak Press, 1950.

Delaware County (NY) Department of Watershed Affairs. *Delaware County Action Plan: Progress Report.* Delhi, NY: Delaware County, June 2006.

Department of Parks of the City of New York. *Altman Memorial Carillon High Bridge Park, New York City.* New York: New York Department of Parks and Recreation, 1958.

———. *Annual Report of the Department of Parks of the City of New York.* New York: Department of Parks, 1905, 1916, 1927.

———. *Improvement and Expansion of New York City Park Facilities.* New York: Department of Parks, 1937.

Department of Water Supply, Gas, and Electricity. *Some of the Things Accomplished by the Department, 1914–1917.* New York: Department of Water Supply, Gas, and Electricity, 1917.

Forstall, Richard. *New York, Population of Counties by Decennial Census: 1900 to 1990.* U.S. Census Bureau website, www.census.gov/population/cencounts/ny190090.txt.

Freud, Salome. *Why New York City Needs a Filtered Croton Supply.* New York: New York City Department of Environmental Protection, March 2003.

Gibson, Campbell. *Population of the 100 Largest Cities and Other Urban Areas in the United States: 1790 to 1990.* Washington, DC: U.S. Bureau of the Census, 1998.

Interstate Commission on the Delaware River Basin. *INCODEL: A Report on Its Activities and Accomplishments.* Philadelphia: Interstate Commission on the Delaware River Basin, 1943.

Long Island State Park Commission. *Annual Report of the Long Island State Parks Commission.* Albany: Long Island State Park Commission. 1925–26.

Mayor's Committee on Management Survey. *Future Water Sources of the City of New York: Report of Engineering Panel on Water Supply to Mayor's Committee on the Management Survey of the City of New York.* New York: Mayor's Committee on Management Survey, 1951.

Mayor's Executive Committee on Administration. *Report to the Mayor Concerning the Water Supply and Immediate Need for Conservation.* New York: Executive Committee on Administration, 1948.

Mayor's Intergovernmental Task Force on New York City Water Supply Needs. *Increasing Supply, Controlling Demand: Interim Report.* New York: New York City Department of Environmental Protection, 1986.

New York City Department of Environmental Protection. *Final Environmental Impact Statement (EIS): The Extended New York City Land Acquisition Program.* New York: Department of Environmental Protection, 2010.

———. *A Handy Resource Guide to the New York City Toilet Rebate Program*, 2nd ed. New York: Department of Environmental Protection, 1992.

———. *New York City Drinking Water Supply and Quality Report.* Corona, NY: Department of Environmental Protection, 1999.

————. *New York City Watershed Memorandum of Agreement.* New York, 1997.

————. *Public Hearings—Croton Water Treatment Plant: Environmental Impact Statements,* volume 5. Hearing held at Lehman College, February 11, 1998. New York: Barrister Reporting Service, 1998.

————. *Water Conservation Program.* New York: Department of Environmental Protection, 2006.

New York City Department of Environmental Protection. Bureau of Water and Energy Conservation. *Water-Saving Plumbing Fixture Retrofit and Rebate Program.* New York: Department of Environmental Protection, 1992.

New York City Department of Environmental Protection. Bureau of Water Supply. *Long-Term Land Acquisition Plan, 2012 to 2022.* New York: Department of Environmental Protection, 2009.

————. *2006 Long-Term Watershed Protection Program.* New York: Department of Environmental Protection, 2006.

New York City Department of Water Supply, Gas and Electricity. *Analysis of Report of Engineering Panel on Water Supply.* New York: Department of Water Supply, 1952.

————. *Some of the Things Accomplished by the Department, 1914–1917.* New York: Department of Water Supply, 1917.

New York State Comptroller. *Delaware Aqueduct System: Water Leak Detection and Repair Program, Report 2005-N-7.* Albany: Comptroller's Office, 2007.

New York State Department of Environmental Conservation. Bureau of Fisheries. Fishery Resources of the Upper Delaware Basin. Albany: Department of Environmental Conservation, January 1975.

New York State Legislature, Assembly Cities Committee. *Brief on Behalf of the Ramapo Water Company in Opposition to Bill, Print No. Nine, Repealing Chapter 985 of the Laws of 1895.* Albany: New York State Assembly, February 1901.

New York State Senate, Committee on the Affairs of the City of New York. *Report on Legislation to Revise Condemnation Procedures Relative to Acquisition of Land in the Delaware River Basin for Water Supply Purposes for the City of New York.* Albany: Committee on the Affairs of the City of New York, 1963.

Sheppard, J. Douglas. *New York Reservoir Releases Monitoring and Evaluation Program—Delaware River—, Technical Report No. 83-5.* Albany: Department of Environmental Conservation, 1983.

State of New Jersey, Complainant, against State of New York and City of New York, Defendants, Commonwealth of Pennsylvania, Intervenor: Before Hon, Kurt F. Pantzer. New York, 1954.

State of New Jersey, Complainant, the Delaware River Case: In the Supreme Court of the United States, October Term, 1929, no. 17, Original / State of New Jersey, Complainant; State of New York and the City of New York, defendant; State of Pennsylvania, Intervener: Before Charles N. Burch, Special Master, April 21 to April 25, 1930, Trenton, NJ. Vol. 4. Washington, DC, 1931.

State Water Supply Commission. *Second Annual Report of the State Water Supply Commission of New York for Year Ending Feb. 1, 1907.* Albany: J. B. Lyon Co., 1907.

Temporary State Commission on the Water Supply Needs of Southeastern New York, *Alternative Futures: A Re-Evaluation.* Albany: Commission on the Water Supply Needs of Southeastern New York, 1974.

U.S. Environmental Protection Agency. *Assessing New York City's Watershed Protection Program: The 1997 Filtration Avoidance Determination Mid-Course Review for the Catskill/Delaware Water Supply Watershed.* New York: Environmental Protection Agency, 2000.

————. *New York City Filtration Avoidance Determination*. New York: Environmental Protection Agency, 2002.

————. *Report on the City of New York's Progress in Implementing the Watershed Protection Program, and Complying with the Filtration Avoidance Determination*. New York: Environmental Protection Agency, 2006.

U.S. Senate Committee on Interior and Insular Affairs. *Northeast Water Crisis: Hearing before the Committee on Interior and Insular Affairs*. Eighty-Ninth Congress, first session, September 8, 1965.

Westchester County Department of Planning. *Databook: Westchester County, New York*. White Plains, NY: Westchester County, 2005.

Westchester County Water Agency. *1976 Annual Report*. White Plains, NY: Westchester County, 1977.

Works Progress Administration. *Historical Development of Westchester County: A Chronology*. Vol. 2. White Plains, NY: Westchester County Emergency Work Bureau, 1939.

————. *Survey of Water Consumption in the City of New York: Final Report, December 1940*. New York: Works Progress Administration and New York City Department of Water Supply, Gas, and Electricity, 1940.

Interviews

Behind the Scenes: The Inside Story of the Watershed Agreement (BTS)

Interviews, transcripts available at www.cwconline.org/about/scenes.html.
Jeffrey S. Baker. Interview by Nancy Burnett. December 15, 1998.
Brooks, Clayton. Interview by Nancy Burnett and Virginia Scheer. January 26, 1996.
Bucca, Anthony. Interview by Nancy Burnett and Virginia Scheer. February 15, 1996.
Crotty, Erin. Interview by Nancy Burnett. August 22, 2001.
Gelber, Marilyn. Interview by Nancy Burnett. July 6, 2001.
Greenfield, Eric. Interview by Nancy Burnett. February 20, 1996.
Kennedy, Robert F., Jr. Interview by Nancy Burnett. July 12, 2002.
Markert, Kenneth. Interview by Nancy Burnett and Virginia Scheer. January 18, 1996.
Porter, Keith. Interview by Nancy Burnett. July 20, 2001.
Rosa, Alan. Interview by Nancy Burnett and Virginia Scheer. February 6, 1996.
Ruzow, Daniel. Interview by Nancy Burnett. August 2, 2001.
Shelton, Perry. Interview by Nancy Burnett. February 20, 1996.

Interviews with the Author

Appleton, Albert. April 9, 2008.
Baker, Jeff. May 26, 2011.
Borrelli, Peter. February 15, 2006.
Collins, Tara, and Fred Huneke. May 24, 2011.
Coluccio, Vincent. June 7, 2008.
Frazier, Dean, and Tom Hilson. May 26, 2011.
Gleising, Fred, and John Porter. February 23, 2006.
Liebold, Warren. January 18, 2007.
McShane, Jack, and Alan White. February 22, 2006.

Rapp, Dennis. February 20, 2008.
Rosa, Alan. May 25, 2011.
Sheppard, J. Douglas. July 8, 2008.
Serio, Jim. April 16, 2008.
Snow, Thomas. August 25, 2008.
Van Valkenburgh, Norm. April 15, 2008.
Warne, David. January 13, 2009.

Books, Pamphlets, Presentations, Articles, and Other Publications

Abbott, Ernest Hamlin. "A Subway for Water." *Outlook*, January 23, 1909, 157–74.
Allen, Edward L., James C. Duane, and Alphonse Fteley. *Report to the Aqueduct Commissioners*. New York: Douglas & Taylor Co., 1895.
Alpern, Robert. *Thirsty City: A Plan of Action for New York City Water Supply*. New York: Citizens Union Foundation, 1986.
American City Magazine. A Report on the New York City Water Supply and the Effect Water Meters Would Have on That Supply. New York: American City Magazine, 1950.
Appleton, Albert. "Financial Tools, Financial Strategy and Water Utility Operating and Environmental Performance—Some Experiences of the New York City Water and Sewer System." Paper presented in China, 2002.
———. "How New York City Used an Ecosystem Services Strategy Carried Out through an Urban-Rural Partnership to Preserve the Pristine Quality of Its Drinking Water and Save Billions of Dollars and What Lessons It Teaches about Using Ecosystem Services." Paper presented at Katoomba Conference, on Forest Trends, Tokyo, November 2002.
———. "Regional Landscape Preservation and the New York City Watershed Protection Program: Some Reflections." Paper presented at "The Race for Space: The Politics and Economics of State Open Space Programs," Princeton, NJ. December 2, 2005.
Barnard, Charles. "The New Croton Aqueduct." *Century Illustrated Magazine*, December 1889, 205–24.
Bond, M. C. *Ulster County: Census Data*. Ithaca, NY: New York State College of Agriculture, 1953.
Borrelli, Peter, ed. *The Catskill Center Plan: A Plan for the Future Conservation and Development of the Catskill Region*. Hobart, NY: Catskill Center for Conservation and Development, 1974.
"Brodhead's and Ashokan Reservoir." *Olde Ulster: An Historical and Genealogical Magazine*, April 1913, 103–5.
Burr, William, Rudolph Hering, and John R. Freeman. *Report of the Commission on Additional Water Supply for the City of New York Made to Robert Grier Monroe, Commissioner of the Department of Water Supply, Gas, and Electricity*. New York: Martin B. Brown Press, 1904.
Butler, George B. *In Favor of Retaining the Present Murray Hill Reservoir*. New York: New York Municipal Society, 1878.
Carpenter, Charles. "Report to the State Forest Commission on the Catskill Preserve." Originally published in the *Second Annual Report of the Forest Commission*. Albany: Forest Commission, 1886.
Catskill Enterprise. *Picturesque Catskill*. Saugerties, NY: Catskill Enterprise, 1932.
Catskill Watershed Corporation. *2009 Annual Report*. Margaretville, NY: Catskill Watershed Corp., 2010.

———. *Summary Guide to the Terms of the Watershed Agreement: A Guidebook for Government Officials, Planning and Zoning Board Members, and Citizens of the Catskill/Delaware Watershed.* Arkville, NY: Catskill Watershed Corp., 1997. (Available at www.cwconline. org/pubs/moa.html.)

Citizens Budget Commission. *Action on Water: The Samuel J. and Ethel Lefrak Foundation, Inc. Report.* New York: Citizens Budget Commission, 1965.

Citizens Union, Committee on City Affairs, Subcommittee on Water Meters. *Water Meters for New York City's Multiple Dwellings: A Discussion of the Department of Water Supply, Gas & Electricity Proposal to Meter All Dwellings Housing Three or More Families.* New York: Citizens Union, 1949.

City Club of New York. *Why the Citizens of New York Should Resist and Defeat the Ramapo Water Scheme.* 1897. Reprint, New York: City Club, 1899.

"Clearing and Grubbing the Ashokan Reservoir." *Catskill Water System News*, March 20, 1913, 169–70.

Cole, Edward. "Water Waste and Its Detection." *Journal of the Society of Western Engineers* 7 (December 1902): 574–95.

Davidson, Maurice. "The Logic of Universal Metering." *American City Magazine*, May 1935, 63–64.

Downeast Development Consulting Group. The New York City Watershed Economic Impact Assessment Report: Determining Impacts and Developing Options regarding NYC's Land Acquisition Program in Delaware County. Downeast Development Consulting Group, May 2009.

Eastman, H. G. *The Water Supply of New York: Suggestions for a Permanent and Economical Settlement of the Question: The Hudson River as an Available, Unlimited, and Unfailing Source: Plan for the Relief of the City before Another Season.* New York: S. W. Green, Printer, 1876.

Eisenhut, L. C. *The Brooklyn Water Supply under the Probe of the Public Searchlight: The Long Island Watershed and Water Supply Engineering.* Brooklyn, 1903.

Fitzgerald, Desmond. "Journey to Catskill Supply, New York Water Works." *Journal of the New England Water Works Association* 32, no. 4 (December 1918): 391–97.

Franko, Alfred M. *Your Mount Vernon Board of Water Supply: An Historical Account of a Community's Struggle for a Pure Water Supply with a Description of the Present System.* Mount Vernon, NY: Mount Vernon Board of Water Supply, 1959.

Freeman, John Ripley. *Report upon New York's Water Supply, with Particular Reference to the Need of Procuring Additional Sources and Their Probable Cost, with Works Constructed under Municipal Ownership, Made to Bird S. Coler, Comptroller.* New York: Martin B. Brown Press, 1900.

Greater Philadelphia Movement. *A Position Paper on Delaware River Water.* Philadelphia: Greater Philadelphia Movement, 1965.

Groopman, Abraham. "Effect of the Northeast Water Crisis on the New York City Water Supply System." *Journal of the American Water Works Association* 60 (January 1968): 37–47.

Hall, Edward Hagaman. *The Catskill Aqueduct and Earlier Water Supplies of the City of New York.* New York: Mayor's Catskill Aqueduct Celebration Committee, 1917.

Hamilton, Rabinovitz & Alschuler Inc. *West of Hudson Development Study for the Catskill Watershed Corporation, Preliminary Draft Report: A Blueprint for the Catskill Fund for the Future.* New York: Hamilton, Rabinovitz & Alschuler, 1998.

Hazen and Sawyer. *Final Impact Assessment Report: Impact Assessment of Natural Gas Produc-tion in the New York City Water Supply Watershed.* New York: New York City Department of Environmental Protection, 2009.

———. *Study of Water Demands on New York City System: Final Report.* New York: New York State Department of Environmental Conservation and New York City Depart-ment of Environmental Protection, 1989.

Heacox, Cecil. "Liquid Assets." *New York State Conservationist,* June–July 1950, 2–9.

Herschel, Clemens, Francis Pruyn, and J. Edmund Woodman. *Report on Distribution of Catskill Water Supply in New York City to a Committee of the Board of Estimate and Ap-portionment.* New York: Martin B. Brown Press, 1910.

Higgins, Frank. *Public Papers of Frank W. Higgins, Governor, 1905.* Albany: J. B. Lyon Co., 1906.

"High Bridge." *Harper's Weekly,* August 22, 1885, 555.

Hoffman, A. W. "Ulster County Bluestone." In Richard Lionel De Lisser, *Picturesque Ulster: A Pictorial Work on the County of Ulster, State of New York.* 1896–1905, in eight parts. Re-print, Cornwallville, NY: Hope Farm Press, 1968.

"Holiday Excursions." *Harper's Weekly,* July 24, 1880, 471.

"International Visitors Learn from Watershed Farmsteads." *Watershed Farm and Forest,* Winter 2009, 1–2.

Irwin, H. G. "Notes on Consumption." *Delaware Water Supply News,* June 15, 1939, 97–98.

———. "Water Consumption: Sanitary Facilities." *Delaware Water Supply News,* August 15, 1939, 116.

Jackson, Daniel Dana. *Pollution of New York Harbor as a Menace to Health by the Dissem-ination of Intestinal Diseases through the Agency of the Common House Fly.* New York: Merchants' Association, 1907.

Kudish, Michael. *The Catskill Forest: A History.* Fleischmanns, NY: Purple Mountain Press, 2000.

Kunz, George Frederick. *Catskill Aqueduct Celebration Publications.* New York: Mayor's Catskill Aqueduct Celebration Committee, 1917.

La Guardia, Fiorello. "Ten Misconceptions of New York." In *Empire City: New York through the Centuries,* edited by Kenneth T. Jackson and David S. Dunbar, 647–52. New York: Columbia University Press, 2002.

Lane, Merritt. *Argument of Merritt Lane in Opposition to the Tri-State Treaty at the Hearing at the State House, Trenton, New Jersey, March 4th, 1927, Supplemented to Answer Certain of the Contentions of the Proponents of the Treaty and Certain Attacks Made on the Argument of the Objectors.* Trenton: Merritt Lane, 1927.

Liebold, Warren. "Conservation Program Development in New York City." Paper present-ed at American Water Works Association Meeting, February 2007.

Longstreth, Thomas Morris. *The Catskills.* 1918. Reprint, Hensonville, NY: Black Dome Press, 2003.

Malcolm Pirnie Engineers. *Report on the Utilization of the Waters of the Delaware River Basin.* Philadelphia: Interstate Commission on the Delaware River Basin, 1950.

Merchants' Association of New York. *An Inquiry into the Conditions Relating to the Water-Supply of the City of New York.* New York: Isaac Blanchard Co., 1900.

———. *Attitude of Candidates for the Legislature toward the Ramapo Water Monopoly: State-ments in Response to Request of the Merchants' Association of New York.* New York: Mer-chants' Association of New York, 1900.

Metz, Herman. *Lands Taken in Condemnation for the New Catskill Water Supply: Their Cost and the Attendant Expenses.* New York: Martin B. Brown Press, 1909.

Morris, George Pope. *The Croton Ode: Written at the Request of the Corporation of the City of New York.* New York: Nesbitt Printer, 1842.

Nevins, Allan, ed. *The Diary of Philip Hone, 1828–1851.* New York: Dodd, Mead, 1936.

New York Academy of Medicine. *The Sewage Disposal Problem of New York City: A Report of the Public Health Committee of the New York Academy of Medicine.* New York: William Wood & Co., 1918.

Nimsgern, E. G. *Illustrated and Descriptive Account of the Main Dams and Dikes of the Ashokan Reservoir.* Brown Station, NY: W. W. Wright, 1909.

O'Neil, Paul. "In Praise of Trout—and Also Me." *Life,* May 8, 1964, 102–18.

Pantzer, Kurt F. *State of New Jersey, Complainant, vs. State of New York, and City of New York, Defendants, Commonwealth of Pennsylvania, and State of Delaware, Intervenors. Report of the Special Master Recommending Amended Decree.* New York: Appeal Printing Co., 1954.

Parrott, R. D. A. "The Water Supply for New York City." *Scientific American Supplement,* September 4, 1886, 8890–92.

Perlee, Brian. *The Future Cost of Water in New York City.* New York: Citizens Budget Commission, 1994.

"Personals." *Catskill Water System News,* August 5, 1911, 16.

"Personals." *Catskill Water System News,* November 5, 1912, 136.

Provost, Andrew. "Protection of New York's Water Supply from Pollution during Construction Work." *Journal of the New England Water Works Association* 25 (September 1911): 301–16.

Public Works Historical Society. *An Interview with Samuel S. Baxter.* Chicago: Public Works Historical Society, 1982.

"The Removal of Bodies from the Ashokan Cemeteries." *Catskill Water System News,* May 20, 1912, 89–90.

"Reservoir Field Day." *Catskill Water System News,* August 5, 1913, 206.

Ridgway, Robert. "The Hudson River Crossing of the Catskill Aqueduct." *Journal of the New England Water Works Association* 25, no. 3 (September 1911): 317–44.

———. *Robert Ridgway.* New York: J. J. Little and Co., 1940.

Rosa, Alan. "Executive Director's Report." *Watershed Advocate,* Spring 1999, 2.

Smyth, Clifford. "New York's Great New Library." *Munsey's Magazine,* February 1906, 516–25.

Stradling, David. *Conservation in the Progressive Era: Classic Texts.* Seattle: University of Washington Press, 2004.

Stroud Water Research Center. *Water Quality Monitoring in the Source Water Areas for New York City: An Integrative Watershed Approach, a Final Report on Monitoring Activities, 2000–2005.* Avondale, PA: Stroud Water Research Center, 2008.

Syrett, Harold C., ed. *The Gentleman and the Tiger: The Autobiography of George B. McClellan, Jr.* New York: J. B. Lippincott Co., 1956.

Ulster and Delaware Railroad. *The Catskill Mountains: The Most Picturesque Mountain Region on the Globe, Ulster and Delaware Railroad.* Rondout, NY: Press of Kingston Freeman, 1902.

Waring, George., Jr., "The Water Famine in Brooklyn." *Harper's Weekly,* February 6, 1897, 134.

Westchester Water Works Conference, Water Resources Committee. *Report on the Water Resources of Westchester County, New York.* White Plains, NY: Westchester Water Works Conference, 1951.

White, Lazarus. *The Catskill Water Supply of New York City: History, Location, Sub-surface Investigations and Construction.* New York: John Wiley & Sons, 1913.

Whitman, Walt. "Murray Hill Reservoir, November 25, 1849." In *Empire City: New York through the Centuries*, edited by Kenneth T. Jackson and David S. Dunbar, 206–8. New York: Columbia University Press, 2002.

Wurster, F. W. The Last Municipal Administration of Brooklyn: Review of the Work Accomplished under Mayor F. W. Wurster. Brooklyn: Eagle Book Printing Dept., 1897.

Secondary Sources

Books, Articles, Book Chapters, and Presentations

Abu-Lughod, Janet L. *New York, Chicago, Los Angeles: America's Global Cities.* Minneapolis: University of Minnesota Press, 1999.

Albert, Richard. *Damming the Delaware: The Rise and Fall of Tocks Island Dam.* State College: Pennsylvania State University Press, 1987.

Alcamo, Joseph, and Elena Bennett. *Ecosystems and Human Well-Being: A Framework for Assessment.* Washington, DC: Island Press, 2003.

"All for Water." *Watershed Advocate*, Summer 2002, 5.

Andersen, Tom. *This Fine Piece of Water: An Environmental History of Long Island Sound.* New Haven, CT: Yale University Press, 2002.

Andrews, Richard N. L. *Managing the Environment, Managing Ourselves: A History of American Environmental Policy.* 2nd ed. New Haven, CT: Yale University Press, 2006.

Appleton, Albert F. "The Post-Delaware Era." In *Water-Works: The Architecture and Engineering of the New York City Water Supply*, edited by Kevin Bone and Gina Pollara, 216–33. New York: Monacelli Press, 2006.

——. "Regional Landscape Preservation and the New York City Watershed Protection Program: Some Reflections." Paper presented at "The Race for Space: The Politics and Economics of State Open Space Programs," Princeton, NJ, December 2, 2005.

Armsworth, P. R., et al. "Ecosystem-Service Science and the Way Forward for Conservation." *Conservation Biology* 21, no. 6 (December 2007): 1383–84.

Aronson, David I. "The City Club of New York, 1892–1912." PhD diss., New York University, 1975.

Arthur, J. Daniel, Brian Bohm, and Mark Layne. "Hydraulic Fracturing Considerations for Natural Gas Wells of the Marcellus Shale." Presented at the Ground Water Protection Council Annual Forum, Cincinnati, September 2008. Available at www.dec.ny.gov/docs/materials_minerals_pdf/GWPCMarcellus.pdf.

Babbitt, Bruce E. *Cities in the Wilderness: A New Vision of Land Use in America.* Covelo, CA: Island Press, 2005.

——. "Diversity: Noah's Mandate and the Birth of Urban Bioplanning." *Conservation Biology* 13 (June 1999): 677–78.

Ballon, Hilary, and Kenneth T. Jackson, eds. *Robert Moses and the Modern City: The Transformation of New York*. New York: W. W. Norton, 2007.

Bateman, Christopher. "A Colossal Fracking Mess." *Vanity Fair*, June 21, 2010. www.vanityfair.com/business/features/2010/06/fracking-in-pennsylvania-201006?currentPage=1.

Bateman, Mark, Frances H. Irwin, and Janet Ranganathan. *Restoring Nature's Capital: An Action Agenda to Sustain Ecosystem Services*. Washington, DC: World Resources Institute, 2007.

Bender, Thomas. *New York Intellect: A History of Intellectual Life in New York City, from 1750 to the Beginnings of Our Own Time*. Baltimore: Johns Hopkins University Press, 1987.

Berman, Marshall. *All That Is Solid Melts into Air: The Experience of Modernity*. New York: Simon & Schuster, 1982.

Birkland, Thomas, et al. "Environmental Policy in New York." In *Governing New York State*, 5th ed., edited by Robert Pecorella and Jeffrey Stonecash, 391–407. Albany: SUNY Press, 2006.

Blackmar, Elizabeth, and Roy Rosenzweig. *The Park and the People: A History of Central Park*. Ithaca, NY: Cornell University Press, 1992.

Blake, Nelson. *Water for the Cities: A History of the Urban Water Supply Problem in the United States*. Syracuse, NY: Syracuse University Press, 1956.

Blakelock, Chester. *History of the Long Island State Parks*. Amityville, NY: Long Island Forum, 1959.

Bloom, Nicholas Dagen. *Public Housing That Worked: New York in the Twentieth Century*. Philadelphia: University of Pennsylvania Press, 2008.

Blumstein, Alfred, and Joel Wallman. *The Crime Drop in America*. New York: Cambridge University Press, 2000.

Bone, Kevin, and Gina Pollara, eds. *Water-Works: The Architecture and Engineering of the New York City Water Supply*. New York: Monacelli Press, 2006.

Boyer, M. Christine. *Manhattan Manners: Architecture and Style, 1850–1900*. New York: Rizzoli, 1985.

Bradshaw, T. M. "New Life for Old Visions." *Catskill Mountain Region Guide*, January 2007. www.catskillmtn.org/guide-magazine/articles/2007-01-new-life-for-old-visions.html.

Brechin, Gary. *Imperial San Francisco: Urban Power, Earthly Ruin*. Berkeley and Los Angeles: University of California Press, 1999.

Brooks, David, and Oliver Brandes. "Why a Water Soft Path, Why Now and What Then?" *International Journal of Water Resources Development* 27 (June 2011): 315–44.

Brunner, Ronald, Toddi Steelman, Lindy Coe-Juell, Christina Cromley, Christine Edwards, and Donna Tucker, eds. *Adaptive Governance: Integrating Science, Policy, and Decision Making*. New York: Columbia University Press, 2005.

Cannato, Vincent. *The Ungovernable City: John Lindsay and His Struggle to Save New York*. New York: Basic Books, 2001.

Canning, Jeff. "Westchester County since World War II: A Changing People in a Changing Landscape." In *Westchester County: The Past Hundred Years, 1883–1983*, edited by Marilyn Weigold, 188–241. Valhalla, NY: Westchester County Historical Society, 1984.

Capossela, Jim. *Good Fishing in the Catskills*. Tarrytown, NY: Northeast Sportsman's Press, 1989.

Carle, David. *Water and the California Dream: Choices for the New Millennium*. San Francisco: Sierra Club Books, 2003.

Caro, Robert. *The Power Broker: Robert Moses and the Fall of New York*. New York: Alfred A. Knopf, 1974.

Carrott, Richard. *The Egyptian Revival: Its Sources, Monuments, and Meaning, 1808–1858*. Berkeley and Los Angeles: University of California Press, 1978.

Cerillo, Augutus, Jr. "The Impact of Reform Ideology: Early Twentieth Century Municipal Government in New York City." In *The Age of Urban Reform: New Perspectives on the Progressive Era*, edited by Michael Ebner and Eugene Tobin, 68–85. Port Washington, NY: Kennikat Press, 1977.

Cohen, Nancy. *The Reconstruction of American Liberalism, 1865–1914*. Chapel Hill: University of North Carolina Press, 2002.

Commoner, Barry. *The Closing Circle: Nature, Man, and Technology*. New York: Alfred A. Knopf, 1971.

Condran, Gretchen A., Henry Williams, and Rose A. Cheney. "The Decline in Mortality in Philadelphia from 1870 to 1930: The Role of Municipal Services." *Pennsylvania Magazine of History and Biography* 108 (April 1984): 153–77.

Crane, Diana. *The Transformation of the Avant-Garde: The New York Art World, 1940–1985*. Chicago: University of Chicago Press, 1997.

Cronin, John, and Robert F. Kennedy Jr. *The Riverkeepers: Two Activists Fight to Reclaim Our Environment as a Basic Human Right*. New Yorker: Scribner, 1997.

Cronon, William. *Changes in the Land*. New York: Hill & Wang, 1983.

———. *Nature's Metropolis: Chicago and the Great West*. New York: W. W. Norton, 1991.

Daily, Gretchen C., and Katherine Ellison. *The New Economy of Nature: The Quest to Make Conservation Profitable*. Washington, DC: Island Press, 2002.

Dawley, Alan. *Struggles for Justice: Social Responsibility and the Liberal State*. Cambridge, MA: Belknap Press of Harvard University Press, 1991.

Diefendorf, Jeffry M., and Kurk Dorsey. "Challenges for Environmental History." In *City, Country, Empire: Landscapes in Environmental History*, edited by Jeffry M. Diefendorf and Kurk Dorsey, 1–9. Pittsburgh: University of Pittsburgh Press, 2005.

Diner, Steven. *A Very Different Age: Americans of the Progressive Era*. New York: Hill & Wang, 1998.

Doctorow, E. L. *Waterworks*. New York: Random House, 1994.

Doig, Jameson. *Empire on the Hudson: Entrepreneurial Vision and Political Power at the Port of New York Authority*. New York: Columbia University Press, 2001.

Dombeck, Michael P., Christopher A. Wood, and Jack E. Williams. "Focus: Restoring Watersheds, Rebuilding Communities." *American Forests* 103 (Winter 1998): 26.

DuMond, Frank L. *Walking through Yesterday in Old West Hurley*. Kalamazoo, MI: Different Dimensions, 1988.

Eisenach, Eldon. *The Lost Promise of Progressivism*. Lawrence: University Press of Kansas, 1994.

Elkind, Sarah. *Bay Cities and Water Politics: The Battle for Resources in Boston and Oakland*. Lawrence: University Press of Kansas, 1998.

Environmental Law Institute. *NEPA Success Stories: Celebrating 40 Years of Transparency and Open Government*. Washington, DC: ELI, 2010.

Evans, Peter B., Dietrich Rueschemeyer, and Theda Skocpol, eds. *Bringing the State Back In*. New York: Cambridge University Press, 1985.

Evers, Alf. *The Catskills: From Wilderness to Woodstock*. Woodstock, NY: Overlook Press, 1982.

Faure, Guy Olivier, and Jeffrey Rubin, eds. *Culture and Negotiation: The Resolution of Water Disputes*. Newbury Park, CA: SAGE, 1993.

Fiege, Mark. "Private Property and the Ecological Commons in the American West." In *Everyday America: Cultural Landscape Studies after J. B. Jackson*, edited by Paul Groth and Chris Wilson, 219–32. Berkeley and Los Angeles: University of California Press, 2003.

Finnegan, Michael. "New York City's Watershed Agreement: A Lesson in Sharing Responsibility." *Pace Environmental Law Review* 14 (Summer 1997): 577–614.

Fitzgerald, Deborah. "Accounting for Change: Farmers and the Modernizing State." In *The Countryside in the Age of the Modern State: Political Histories of Rural America*, edited by Robert D. Johnston and Catherine McNicol Stock, 189–212. Ithaca, NY: Cornell University Press, 2001.

Flippen, J. Brooks. *Nixon and the Environment*. Albuquerque: University of New Mexico Press, 2000.

Francis, Austin McK. *Land of Little Rivers: A Story in Photos of Catskill Fly Fishing*. New York: Beaverkill Press, 1999.

Freeman, Joshua. *Working-Class New York: Life and Labor since World War II*. New York: Free Press, 2000.

Freyer, Frederic Q. "The Consolidation of New York and Brooklyn: A View of the Trend toward Unification of the Two Cities, 1830–1898." BA thesis, Harvard College, 1983.

"Friendlier Signs." *Watershed Advocate*, Fall 2000, 7.

"From Frustration to Joy: Amy Jackson's Triumph." *Watershed Advocate*, Summer 2006, 1–2.

Galusha, Diane. *Liquid Assets: A History of New York City's Water System*. 1999. Reprint with minor corrections, Fleischmanns, NY: Purple Mountain Press, 2002.

Gamber, Wendy. *The Boardinghouse in Nineteenth-Century America*. Baltimore: Johns Hopkins University Press, 2007.

Gill, Brendan. "The Gathered Waters." *New Yorker*, December 31, 1949, 24–37.

Gleick, Peter. "The Changing Water Paradigm: A Look at Twenty-First Century Water Resources Development." *Water International* 25 (March 2000): 127–38.

Goldman, Joanne Abel. *Building New York's Sewers: The Evolution of Mechanisms of Urban Development*. West Lafayette, IN: Purdue University Press, 1997.

Goldstein, Eric A., and Robin A. Marx. *Under Attack: New York's Kensico and West Branch Reservoirs Confront Intensified Development*. New York: Natural Resources Defense Council and Federated Conservationists of Westchester County, 1999.

"Gone, but Not Forgotten." *Watershed Advocate*, Summer 2004, 8.

Gordon, David, and Robert F. Kennedy Jr. *The Legend of City Water: Recommendations for Rescuing the New York City Water Supply*. Garrison, NY: Hudson Riverkeeper Fund, 1991.

Gordon, Jim. "City Tunnel, Catskill Water." *Ulster: A Regional Magazine*, Summer 1987, 120–33.

Gottehrer, Barry. *New York City in Crisis: A Study in Depth of Urban Sickness*. New York: Pocket Books, 1965.

Grann, David. "City of Water." *New Yorker*, September 1, 2003, 88–109.

Grant, Peter. "Development Surge Upstate a Concern for City's Water." *New York Observer*, November 16, 1987.

Gutfreund, Owen. "Rebuilding New York in the Auto Age: Robert Moses and His Highways." In *Robert Moses and the Modern City: The Transformation of New York*, edited by Hilary Ballon and Kenneth Jackson, 86–93. New York: W. W. Norton, 2007.

Gutman, Marta. "Equipping the Public Realm: Rethinking Robert Moses and Recreation." In *Robert Moses and the Modern City: The Transformation of New York*, edited by Hilary Ballon and Kenneth Jackson, 72–85. New York: W. W. Norton, 2007.

Hanover, Crane. *Guide to Trout Streams in the Catskill Mountains*. Ithaca, NY: Outdoor Publications, 1968.

Hayden, Dolores. "What Is Suburbia? Naming the Layers in the Landscape." In *Westchester: The American Suburb*, edited by Roger Panetta, 77–102. New York: Fordham University Press, 2006.

Hays, Samuel. *Beauty, Health, and Permanence: Environmental Politics in the United States, 1955–1985*. New York: Cambridge University Press, 1987.

Hays, Samuel. "Theoretical Implications of Recent Work in the History of American Society and Politics." *History and Theory* 26 (February 1987): 15–31.

Hirshleifer, Jack, James C. De Haven, and Jerome W. Milliman. *Water Supply: Economics, Technology, and Policy*. Chicago: University of Chicago Press, 1960.

Hogarty, Richard. *The Delaware River Drought Emergency*. New York: Bobbs-Merrill, 1969.

Hood, Clifton. *722 Miles: The Building of the Subways and How They Transformed New York*. New York: Simon & Schuster, 1993.

Howarth, Robert W., Renee Santoro, and Anthony Ingraffea. "Methane and the Greenhouse-Gas Footprint of Natural Gas from Shale Formations." *Climatic Change Letters* 106 (April 2011): 679–90.

Howe, Allen M. *Old West Hurley Revisited: A Nostalgic Tour*. Saugerties, NY: Hope Farm Press, 1999.

Hudson River Museum. *The Old Croton Aqueduct: Rural Resources Meet Urban Needs*. Yonkers, NY: Hudson River Museum, 1992.

Hufschmidt, Maynard. "The Supreme Court and Interstate Water Problems: The Delaware Basin Example." Government 224A. Rev. ed., 1958. Photocopy.

Hundley, Norris. *The Great Thirst: Californians and Water: A History*. 1992. Reprint, Berkeley and Los Angeles: University of California Press, 2001.

"International Visitors Learn from Watershed Farmsteads." *Watershed Farm and Forest*, Winter 2009, 1–2.

Jacobs, Jane. *The Death and Life of Great American Cities*. New York: Random House, 1961.

Jacobs, Meg, William J. Novak, and Julian E. Zelizer, eds. *The Democratic Experiment: New Directions in American Political History*. Princeton, NJ: Princeton University Press, 2003.

Jacoby, Karl. *Crimes against Nature: Squatters, Poachers, Thieves, and the Hidden History of American Conservation*. Berkeley and Los Angeles: University of California Press, 2001.

John, DeWitt. *Civic Environmentalism: Alternatives to Regulation in States and Communities*. Washington, DC: CQ Press, 1994.

Josephson, Paul. *Resources under Regimes: Technology, Environment, and the State*. Cambridge, MA: Harvard University Press, 2004.

Judd, Peter H. *How Much Is Enough? Controlling Water Demand in Apartment Buildings*. Denver: American Water Works Association, 1993.

Kahrl, William. *Water and Power: The Conflict over Los Angeles' Water Supply in the Owens Valley*. Berkeley and Los Angeles: University of California Press, 1982.

Kamieniecki, Sheldon, and Michael E. Kraft, eds. *Business and Environmental Policy: Corporate Interests in the American Political System*. Cambridge, MA: MIT Press, 2007.

Kaplan, Richard. *Politics and the American Press: The Rise of Objectivity, 1865–1920*. New York: Cambridge University Press, 2002.

Kelman, Ari. *A River and Its City: The Nature of Landscape in New Orleans*. Berkeley and Los Angeles: University of California Press, 2003.

Kennedy, Robert F., Jr., Jeffrey Odefey, William Wegner, and Marc Yaggi. *Finger in the Dike, Head in the Sand: DEP's Crumbling Water Supply Infrastructure*. Elmsford, NY: Riverkeeper, 2001.

Kennedy, Robert F., Jr., Mark Sullivan, and Mary Beth Postman. *Watershed for Sale: Explosive Development Threatens New York City's Water Supply*. New York: Riverkeeper, 1999.

Keyes, Jonathan. "A Place of Its Own: Urban Environmental History." *Journal of Urban History* 26, no. 3 (March 2000): 380–90.

Kingsland, Sharon. *The Evolution of American Ecology, 1890–2000*. Baltimore: Johns Hopkins University Press, 2005.

Kinkela, David. "The Ecological Landscapes of Jane Jacobs and Rachel Carson." *American Quarterly* 61 (December 2009): 905–28.

Klein, Jennifer. *For All These Rights: Business, Labor, and the Shaping of America's Public-Private Welfare State*. Princeton, NJ: Princeton University Press, 2003.

Klingle, Matthew. *Emerald City: An Environmental History of Seattle*. New Haven, CT: Yale University Press, 2007.

Klyza, Christopher McGrory, and David Sousa. *American Environmental Policy, 1990–2006: Beyond Gridlock*. Cambridge, MA: MIT Press, 2008.

Koeppel, Gerard T. "The Rise to Croton." In *Water-Works: The Architecture and Engineering of the New York City Water Supply*, edited by Kevin Bone and Gina Pollara, 26–51. New York: Monacelli Press, 2006.

———. *Water for Gotham: A History*. Princeton, NJ: Princeton University Press, 2000.

Krieg, Joann. *Robert Moses: Single-Minded Genius*. Interlaken, NY: Heart of Lakes Press, 1989.

Layzer, Judith. *Natural Experiments: Ecosystem-Based Management and the Environment*. Cambridge, MA: MIT Press, 2008.

Levine, Steven A. "In Gotham's Shadow: Brooklyn and the Consolidation of Greater New York." PhD diss., City University of New York, 2002.

Logan, Daniel. "The Ashokan: A Winter View." *Catskill Center News*, Winter 1989, 9.

Logan, Michael. *Desert Cities: The Environmental History of Phoenix and Tucson*. Pittsburgh: University of Pittsburgh Press, 2006.

Lubell, Mark, et al. "Watershed Partnerships and the Emergence of Collective Action Institutions." *American Journal of Political Science* 46 (January 2002): 148–63.

Mahler, Jonathan. *Ladies and Gentlemen, the Bronx Is Burning: 1977, Baseball, Politics, and the Battle for the Soul of a City*. New York: Picador, 2005.

Mandarano, Lynn. "Clean Water, Clean City: Sustainable Storm Water Management in Philadelphia." In *Sustainability in America's Cities: Creating the Green Metropolis*, edited by Matthew I. Slavin, 157–79. Washington, DC: Island Press, 2011.

Martin, Roscoe C. *Metropolis in Transition: Local Government Adaptation to Changing Urban Needs*. Washington, DC: Housing and Home Finance Agency, 1963.

———. *Water for New York: A Study in State Administration of Water Resources*. Syracuse, NY: Syracuse University Press, 1960.

Mattera, John. "New York City's Water Parks." *Daily Plant*, September 9, 2004.

McMahon, Jane. "Westchester from the Roaring Twenties to V-J Day." In *Westchester County: The Past Hundred Years, 1883–1983*, edited by Marilyn Weigold, 109–46. Valhalla, NY: Westchester County Historical Society, 1984.

McPhee, John. *The Control of Nature*. New York: Farrar, Straus and Giroux, 1990.

Melnick, R. Shep. "Risky Business Government and the Environment after Earth Day." In *Taking Stock: American Government in the Twentieth Century*, edited by Morton Keller and R. Shep Melnick, 156–84. New York: Cambridge University Press, 1999.

Melosi, Martin. *Precious Commodity: Providing Water for America's Cities*. Pittsburgh: University of Pittsburgh Press, 2011.

———. *The Sanitary City: Urban Infrastructure in America from Colonial Times to the Present*. Baltimore: Johns Hopkins University Press, 2000.

Milazzo, Paul Charles. *Unlikely Environmentalists: Congress and Clean Water, 1945–1972*. Lawrence: University Press of Kansas, 2006.

Mitchell, John G. *The Catskills: Land in the Sky*. New York: Viking, 1977.

Mohl, Raymond A. "City and Region: The Missing Dimension in US Urban History." *Journal of Urban History* 25 (November 1998): 3–21.

National Research Council. Committee on Assessing and Valuing the Services of Aquatic and Related Terrestrial Ecosystems, Water Science and Technology Board, Commission on Geosciences, Environment, and Resources. *Valuing Ecosystem Services: Toward Better Environmental Decision-Making*. Washington, DC: National Academy Press, 2004.

———. Committee on Watershed Management. *New Strategies for America's Watersheds*. Washington, DC: National Academy Press, 1999.

———. Committee to Review the New York City Watershed Management Strategy. *Watershed Management for Potable Water Supply: Assessing the New York City Strategy*. Washington, DC: National Academy Press, 2000.

Nesson, Fern. *Great Waters: A History of Boston's Water Supply*. Hanover, NH: University Press of New England, 1983.

Novak, William J. "The Myth of the 'Weak' American State." *American Historical Review* 113, no. 3 (June 2008): 752–72.

Orren, Karen, and Stephen Skowronek. *The Search for American Political Development*. New York: Cambridge University Press, 2004.

Owen, David. *Green Metropolis: Why Living Smaller, Living Closer, and Driving Less Are the Keys to Sustainability*. New York: Riverhead, 2009.

Page, Max. *The Creative Destruction of Manhattan, 1900–1940*. Chicago: University of Chicago Press, 1999.

Panetta, Roger, ed. *Westchester: The American Suburb*. New York: Fordham University Press, 2006.

Perrin, Noel. "New York Drowns Another Valley." *Harper's Magazine*, August 1963, 76–83.

Peterson, Jon. "The Impact of Sanitary Reform upon American Planning." *Journal of Social History* 13, no. 1 (1979): 83–103.

Pires, Mark. "Watershed Protection for a World City: The Case of New York." *Land Use Policy* 21 (April 2004): 161–75.

Pisani, Donald. "Beyond the Hundredth Meridian: Nationalizing the History of Water in the United States." *Environmental History* 5 (October 2000): 466–82.

Platt, Rutherford H. "Water Supply Protection through Watershed Management: The New York City and Metro Boston Strategies." Presentation to the NAE Convocation of Professional Engineering Societies, Public Policy Symposium, Washington, DC, May 16, 2011.

Platt, Rutherford H., Andrew J. Hoffman, and Robin Gregory. "A Full Clean Glass?" *Environment* 42 (June 2000): 8–20.

Platt, Rutherford H., Rowan Rowntree, and Pamela Muick. *The Ecological City: Preserving and Restoring Urban Biodiversity*. Amherst: University of Massachusetts Press, 1994.

Pritchard, Evan T. *Native New Yorkers: The Legacy of the Algonquin People of New York*. Tulsa, OK: Council Oak Books, 2007.

Pugliese, Donato Joseph. "Citizen Group Activity in the Delaware River Basin." PhD diss., Syracuse University, 1960; reprint, Ann Arbor, MI: University Microfilms, 1967.

Rae, Douglas. *City: Urbanism and Its End*. New Haven, CT: Yale University Press, 2003.

"REDD May Yet Survive Copenhagen Failures," Carbon Positive website, December 21, 2009. www.carbonpositive.net/viewarticle.aspx?articleID=1786.

Reed, Henry Hope. *The New York Public Library: Its Architecture and Decoration*. New York: W. W. Norton, 1986.

Renner, James. *Washington Heights, Inwood, and Marble Hill*. Charleston, SC: Arcadia Publishing, 2007.

Righter, Robert W. *The Battle over Hetch Hetchy: America's Most Controversial Dam and the Birth of Modern Environmentalism*. New York: Oxford University Press, 2006.

Rodgers, Daniel T. "In Search of Progressivism." *Reviews in American History* 10 (December 1982): 113–32.

Rogers, Elizabeth Barlow. *Rebuilding Central Park: A Management and Restoration Plan*. Cambridge, MA: MIT Press, 1987.

Rowcroft, Petrina. "Payments for Environmental Services: A Review of Global Experiences and Recommendations for Their Application in the Lower Mekong Basin." *Working Paper no. 17*. Vientiane, Laos: Mekong River Commission for Sustainable Development, 2005.

Rymer, Russ. "Reuniting a River." *National Geographic*, December 2008, 134–55.

Sabatier, Paul, et al., eds. *Swimming Upstream: Collaborative Approaches to Watershed Management*. Cambridge, MA: MIT Press, 2005.

Sabatier, Paul, Chris Weible, and Jared Ficker. "Eras of Water Management in the United States: Implications for Collaborative Watershed Approaches." In *Swimming Upstream: Collaborative Approaches to Watershed Management*, edited by Paul Sabatier, Will Focht, Mark Lubell, Zev Trachtenberg, Arnold Vedlitz, and Marty Matlock, 23–52. Cambridge: MA: MIT Press, 2005.

Sagoff, Mark. "On the Value of Natural Ecosystems: The Catskill Parable." *Politics and the Life Sciences* 21 (March 2002): 19–25.

Schultz, Stanley K., and Clay McShane. "To Engineer the Metropolis: Sewers, Sanitation, and City Planning in Late-Nineteenth Century America." *Journal of American History* 65, no. 2 (September 1978): 389–411.

Schwartz, Anne. "The Gates and the Water Filtration Plant." *Gotham Gazette*, February 18, 2005. www.gothamgazette.com/article/parks/20050218/14/1328.

Schwartz, Joel. *The New York Approach: Robert Moses, Urban Liberals, and Redevelopment of the Inner City*. Columbus: Ohio State University Press, 1993.

Scobey, David M. *Empire City: The Making and Meaning of the New York City Landscape*. Philadelphia: Temple University Press, 2002.

Scott, James. *Seeing Like a State: How Certain Schemes to Improve the Human Condition Have Failed*. New Haven, CT: Yale University Press, 1998.

Seligman, Amanda. "Producing the North American Metropolitan Landscape." *Journal of Urban History* 34, no. 4 (May 2008): 695–703.

Shabecoff, Philip. *A Fierce, Green Fire: The American Environmental Movement*. Rev. ed. Washington, DC: Island Press, 2003.

Sherk, George William. *Dividing the Waters: The Resolution of Interstate Water Conflicts in the United States*. London: Kluwer Law International, 2000.

Shupe, Barbara, Janet Steins, and Jyoti Pandit. *New York State Population, 1790–1980: A Compilation of Federal Census Data*. New York: Neal-Schuman, 1987.

Sickler, Vera Van Steenbergh. *The Town of Olive through the Years*. Kingston, NY, 1976.

———. *The Town of Olive through the Years, Part Two*. Kingston, NY: Smith Printing, 1979.

Siegel, Frederick. *The Prince of the City: Giuliani, New York and the Genius of American Life*. San Francisco: Encounter Books, 2005.

Sinding, Kate, and Eric Goldstein. "Natural Gas Drilling Threatens NYC Water Supply." *Crain's New York Business.com*, July 17, 2011, www.crainsnewyork.com/article/20110717/SUB/307179981.

Skowronek, Stephen. *Building a New American State: The Expansion of National Administrative Capacities, 1877–1920*. New York: Cambridge University Press, 1982.

Slichter, Gertrude A. "Franklin D. Roosevelt's Farm Policy as Governor of New York State, 1928–1932." *Agricultural History* 33 (October 1959): 167–76.

Smiley, Jane. *Catskill Crafts: Artisans of the Catskill Mountains*. New York: Crown, 1988.

Smith, Henry Nash. *Virgin Land: The American West as Symbol and Myth*. Cambridge, MA: Harvard University Press, 1950.

Spann, Edward K. *The New Metropolis: New York City, 1840–1857*. New York: Columbia University Press, 1981.

Spence, Mark David. *Dispossessing the Wilderness: Indian Removal and the Making of the National Parks*. New York: Oxford University Press, 1999.

Steinberg, Philip E., and George E. Clark. "Troubled Water? Acquiescence, Conflict, and the Politics of Place in Watershed Management." *Political Geography* 18, no. 4 (May 1999): 477–508.

Stephenson, Bruce. "Urban Environmental History: The Essence of a Contradiction." *Journal of Urban History* 31, no. 6 (September 2005): 887–98.

Stern, Robert A. M., Gregory Gilmartin, and John Montague Massengale. *New York 1900: Metropolitan Architecture and Urbanism, 1890–1915*. New York: Rizzoli, 1983.

Stern, Robert A. M., Thomas Mellins, and David Fishman. *New York 1880: Architecture and Urbanism in the Gilded Age*. New York: Monacelli Press, 1999.

Steuding, Bob. *Last of the Handmade Dams: The Story of the Ashokan Reservoir*. Rev. ed. Fleischmanns, NY: Purple Mountain Press, 1989.

Stock, Catherine McNichol, and Robert Johnston. Introduction to *The Countryside in the Age of the Modern State: Political Histories of Rural America*, edited by Catherine McNichol Stock and Robert Johnston, 1–12. Ithaca, NY: Cornell University Press, 2001.

Stradling, David. "Bishop Falls, New York City, and the Contested Value of Land." *New York History* 87 (Fall 2006): 401–21.

———. *Making Mountains: New York City and the Catskills*. Seattle: University of Washington Press, 2008.

Stutz, Bruce. *Natural Lives, Modern Times: People and Places of the Delaware River*. New York: Crown, 1992.

Sugrue, Thomas. *Origins of the Urban Crisis: Race and Inequality in Postwar Detroit*. Princeton, NJ: Princeton University Press, 1996.

Sussman, Glen, Byron W. Daynes, and Jonathan P. West. *American Politics and the Environment*. New York: Longman, 2002.

Tarr, Joel. *The Search for the Ultimate Sink: Urban Pollution in Historical Perspective.* Akron, OH: University of Akron Press, 1996.

Teaford, Jon. *Unheralded Triumph: City Government in America, 1870–1900.* Baltimore: Johns Hopkins University Press, 1983.

Tobias, Dave. "Protection of New York City's Water Supply through Land Acquisition and Stewardship." *Water Resources Impact* 1, no. 5 (September 1999): 9–15.

Troesken, Werner. "Water and Urban Development." *Journal of Urban History* 32, no. 4 (May 2006): 619–30.

Troetel, Barbara. "Suburban Transportation Redefined: America's First Parkway." In *Westchester: The American Suburb*, edited by Roger Panetta, 247–90. New York: Fordham University Press, 2006.

United Nations. Economic Commission for Europe. *Recommendations on Payments for Ecosystem Services in Integrated Water Resources Management.* New York: United Nations, 2007.

———. World Water Assessment Program. *Water for People, Water for Life: A Joint Report by the Twenty Three UN Agencies Concerned with Freshwater.* New York: UNESCO Publishing, 2003.

Van Put, Ed. *Trout Fishing in the Catskills.* New York: Skyhorse, 2007.

Van Valkenburgh, Norman, and Christopher Olney. *The Catskill Park: Inside the Blue Line; the Forest Preserve and Mountain Communities of America's First Wilderness.* Hensonville, NY: Black Dome Press, 2004.

Wallace, Mike, and Edmund Burrows. *Gotham: A History of New York City to 1898.* New York: Oxford University Press, 1999.

Walter, M. Todd, and Michael F. Walter. "The New York City Watershed Agricultural Program (WAP): A Model for Comprehensive Planning for Water Quality and Agricultural Economic Viability." *Water Resources Impact* 1, no. 5 (September 1999): 5–8.

Warner, Sam Bass. *Streetcar Suburbs: The Process of Growth in Boston, 1870–1900.* Cambridge, MA: Harvard University Press, 1962.

"Water Conservation Program on a Vast Scale: New York City Fights a Water Shortage with Publicity, Inspections, Cooperation." *American City Magazine*, April 1940, 55–59.

"Water Plant Making Parks Greener." *Metro New York City Parks*, September 2007, 2.

"Water Quality Protection." *Catskill Center News*, Summer 1988, 14.

Weidner, Charles H. *Water for a City: A History of New York City's Problem from the Beginning to the Delaware River System.* New Brunswick, NJ: Rutgers University Press, 1974.

Weigold, Marilyn. *The Long Island Sound: A History of Its People, Places, and Environment.* New York: NYU Press, 2004.

White, Richard. *The Organic Machine.* New York: Hill & Wang, 1995.

Whiteley, John, Helen Ingram, and Richard W. Perry. *Water, Place, and Equity.* Cambridge, MA: MIT Press, 2008.

Wiebe, Robert. *The Search for Order, 1877–1920.* New York: Hill & Wang, 1967.

Wilson, William H. *The City Beautiful Movement: Creating the North American Landscape.* Baltimore: Johns Hopkins University Press, 1989.

Wiltse, Jeff. *Contested Waters: A Social History of Swimming Pools in America.* Chapel Hill: University of North Carolina Press, 2007.

Wondolleck, Julia M., and Steven L. Yaffee. *Making Collaboration Work: Lessons from Innovation in Natural Resource Management.* Washington, DC: Island Press, 2001.

Videocassettes and DVDs

Carey, Tobey, Artie Traum, and Robbie Dupree. *Deep Water: Building the Catskill Water System*. 45 min. Willow Mixed Media, 2001. Videocassette.

Fox, Josh. *Gasland*. HBO Documentaries, 2010. DVD.

Harty, Drew, and Nelson Bradshaw, directors. *Of Streams and Dreams: The Programs of the Catskill Watershed Corporation*. Catskill Watershed Corp., 2007. DVD.

Town of Olive Clerk's Office. *Step Back in Time*. 1997. Videocassette.

Websites

America's Natural Gas Alliance. "Hydraulic Fracturing: How It Works." www.anga.us/learn-the-facts/hydraulic-fracturing-101.

Catskill Watershed Corporation. *Board of Directors Meeting Minutes*, February 2, 2010. www.cwconline.org/about/ab_board.html.

———. "$310,000 in Economic Development Grants Awarded." Press release, June 1, 2004. www.cwconline.org/news/press/2004/2004_0601.htm.

New York City Department of Environmental Protection. "DEP Completes Ashland Wastewater Treatment Plant." Press release, August 18, 2011. www.nyc.gov/html/dep/html/press_releases/11–76pr.shtml.

———. "DEP Expands Access for Kensico and New Croton Reservoirs." Press release, December 1, 2010. www.nyc.gov/html/dep/html/press_releases/10–102pr.shtml.

———. "DEP Plans to Open 6,600 More Watershed Acres for Recreation." Press release, April 29, 2011. www.nyc.gov/html/dep/html/press_releases/11–34pr.shtml.

———. "DEP Unveils Design to Repair Leaks in the 85-Mile Delaware Aqueduct." Press release, November 19, 2010. www.nyc.gov/html/dep/html/press_releases/10–99pr.shtml.

———. "New York City to Open Watershed Lands for Deer Hunting and Establish Bow-Hunting Season; Permits Required." Press release, August 12, 2002. www.nyc.gov/html/dep/html/press_releases/02–35pr.shtml.

———. "NYC to Acquire 1,655 Acres of Land for Watershed Protection." Press release, August 16, 2011. www.nyc.gov/html/dep/html/press_releases/11–75pr.shtml.

———. "Proposed Amended Rules to Expand Recreational Use in Watershed." Press release, August 16, 2008. http://home2.nyc.gov/html/dep/html/press_releases/08–16pr.shtml.

———. "State, City Announce Landmark Agreement to Safeguard New York City Water." Press release, February 16, 2011. www.nyc.gov/html/dep/html/press_releases/11–11pr.shtml.

———. "Wastewater Treatment Plant Upgrades Moving Ahead." Press release, October 2, 2001. www.nyc.gov/html/dep/html/press_releases/01–40pr.shtml.

———. "Waterfowl Management Program." http://home2.nyc.gov/html/dep/html/watershed_protection/html/waterfowl.html.

New York City Parks and Recreation Department. "A Bronx Story: Four Bronx Parks Receive $14 Million in Renovations." Press release, August 14, 2008. www.nycgovparks.org/sub_newsroom/press_releases/press_releases.php?id=20532.

New York State Department of Environmental Conservation. "History of DEC." www.dec.ny.gov/about/9677.html.

———. "SEQR: Environmental Impact Assessment in New York State." www.dec.ny.gov/permits/357.html.

"NY Watershed Towns' Lawsuit Fails." *AWWA Streamlines* 20 (January 20, 2009). www.awwa.org/publications/StreamlinesArticle.cfm?itemnumber=44937.

Riverkeeper. "Memorandum of Support for S6276 and A10140." www.riverkeeper.org/ . . ./RvK-Memorandum-of-Support-A10140-Wawarsing-Flooding-1-pdf.

U.S. Environmental Protection Agency. "EPA Grants NYC New Waiver from Filtering Drinking Water from Its Catskill/Delaware System." EPA press release, July 30, 2007. http://yosemite.epa.gov/opa/admpress.nsf/3881d73f4d4aaa0b85257359003f5348/54aeb32b2719f5f585257328004c70da!OpenDocument.

———. "EPA Order against NYC Points Out Need to Filter Croton Supply." EPA press release, January 24, 2003. http://yosemite.epa.gov/opa/admpress.nsf/89745a330d4ef8b9852572a000651fe1/f8a75d3ba2c41e24852571630061d778!OpenDocument.

Watershed Agriculture Council website. www.nycwatershed.org/ag_planning.html.

Watershed Post, blog. Julia Reischel, "New Tax Scheme at Center of Watershed Deal." November 15, 2010. www.watershedpost.com/2010/new-tax-scheme-center-watershed-deal.

Water-Technology.net. "New York City Tunnel No. 3 Construction, USA." www.water-technology.net/projects/new-york-tunnel-3/.

Index